Withdrawn from
Davidson College Library

Library of
Davidson College

Technical Progress and
Soviet Economic Development

Technical Progress and Soviet Economic Development

edited by
RONALD AMANN
and
JULIAN COOPER

Basil Blackwell

© R. Amann and J. M. Cooper, 1986

First published 1986

Basil Blackwell Ltd
108 Cowley Road,
Oxford OX4 1JF, UK

Basil Blackwell Inc.
432 Park Avenue South,
Suite 1505,
New York, NY 10016, USA

All rights reserved. Except for the quotation of short passages for the purposes of criticism and review, no part of this publication may be reproduced, stored in a retrieval system, or transmitted, in any form or by any means, electronic, mechanical, photocopying, recording or otherwise, without the prior permission of the publisher.

Except in the United States of America, this book is sold subject to the condition that it shall not, by way of trade or otherwise, be lent, re-sold, hired out, or otherwise circulated without the publisher's prior consent in any form of binding or cover other than that in which it is published and without a similar condition including this condition being imposed on the subsequent purchaser.

British Library Cataloguing in Publication Data

Technical progress and Soviet economic development.
 1. Technological innovations—Economic aspects—
Soviet Union
 I. Amann, Ronald II. Cooper, Julian, *1945–*
338'.06'0947   HC340.T4
ISBN 0-631-14572-9

Library of Congress Cataloging in Publication Data

Main entry under title:
Technical progress and Soviet economic development.
 Includes index.
  1. Soviet Union—Industries—1957– —Addresses, essays, lectures. 2. Soviet Union—Economic policy—Addresses, essays, lectures. 3. Technology transfer—Soviet Union—Addresses, essays, lectures. I. Amann, Ronald, 1943– . II. Cooper, Julian, 1945– .
HC336.T35 1986   338.947'06   85-20093
ISBN 0-631-14572-9

Typeset by Katerprint Co. Ltd, Oxford
Printed in Great Britain by The Bath Press, Avon

Contents

List of contributors	vi
List of tables, figures, charts and graphs	vii
Introduction	1

1 RONALD AMANN
  Technical progress and Soviet economic development: setting the scene — 5

2 JULIAN COOPER
  The civilian production of the Soviet defence industry — 31

3 PAUL SNELL
  Soviet microprocessors and microcomputers — 51

4 ANTHONY RIMMINGTON
  Soviet biotechnology: the case of single cell protein — 75

5 MALCOLM R. HILL and RICHARD MCKAY
  Soviet product quality, state standards and technical progress — 94

6 GARY K. BERTSCH
  Technology transfers and technology controls: a synthesis of the Western-Soviet relationship — 115

7 VLADIMIR SOBELL
  Technology flows within Comecon and channels of communication — 135

8 DAVID A. DYKER
  Soviet planning reforms from Andropov to Gorbachev — 153

9 DANIEL L. BOND
  Prospects for the Soviet economy — 170

References — 183

Index — 211

List of Contributors

RONALD AMANN is Professor of Soviet Politics and Director of the Centre for Russian and East European Studies, University of Birmingham, England (CREES).

GARY K. BERTSCH is Sandy Beaver Professor of Political Science at the University of Georgia, USA. During the 1984–5 session he was a Visiting Fulbright Scholar in the Department of Politics, University of Lancaster.

DANIEL L. BOND is a Senior Vice President of Wharton Econometric Forecasting Associates, Washington, DC, USA.

JULIAN COOPER is Lecturer in Soviet Technology and Industry at CREES, University of Birmingham.

DAVID A. DYKER is Lecturer in the Department of Economics and the School of European Studies at the University of Sussex.

MALCOLM R. HILL is a Senior Lecturer in the Department of Management Studies, University of Technology, Loughborough.

RICHARD McKAY is a Research Associate, also in the Department of Management Studies, Loughborough University.

ANTHONY RIMMINGTON is a doctoral student at CREES.

PAUL SNELL is a doctoral student at CREES.

VLADIMIR SOBELL, formerly at St Antony's College, Oxford, is now a researcher with the Czechoslovak Service of Radio Free Europe.

List of Tables

1.1	Adoption of new technologies: dates of first commercial production	12
1.2	Subsequent diffusion of new technologies: patterns of output within commodity categories in 1982	13
1.3	Major reform options for speeding up technical progress in the USSR	22
1A.1	Invention: number of prototypes of new machines, equipment, apparatus and instruments	26
1A.2	Innovation: expenditures on the introduction of new technologies into industry and their economic impact	27
1A.3	Diffusion: modernization of industrial processes in different branches of industry	28
1A.4	Incremental improvements: inventions and rationalization measures introduced into the national economy	29
1A.5	Rates of growth of labour productivity by branches of industry	30
2.1	The share of total Soviet output of civilian products from enterprises of the defence industry	41
3A.1	Soviet microprocessor series	66
3A.2	Summary of Soviet microprocessor families	73
3A.3	Soviet microcomputers	74
4.1	Production of SCP from n-paraffins	79
4.2	Analysis of SCP produced by BVK (n-paraffin) plants	81
4.3	Quantity of infective bacteria in SCP	84
4.4	Microbiological standards of new or unconventional proteins for use in animal feeds (IUPAC recommendations, 1979)	85
4.5	Production of SCP in the USSR	90
5.1	Comparison of accuracy requirements specified by British and Soviet standards for machine-tool alignment tests	105

6.1	Comparison of high-technology exports with manufactured goods and total goods exports – 17 Western industrial countries to the Communist countries and to the world	118
6.2	USSR sources of Western industrial high-technology products	119
6.3	Selected Soviet and East European legal and illegal acquisitions from the West affecting key areas of Soviet military technology	125
7.1	Indices of mutual turnover and exports of machinery and equipment	142
7.2	Shares of machinery and equipment in total exports to CMEA	142
7.3	Shares of specialized exports in total exports by groups of engineering products	143
7.4	Intra-Comecon supplies of machinery and equipment in 1980	145
7.5	International organizations participating in intra-CMEA diffusion of technology	151
7.6	Number of scientific and technical co-ordinating centres	152
8.1	Average annual rates of growth of Soviet national income	153

List of figures

9.1	Growth of GNP by sector	172
9.2	Industrial and infrastructural growth	173
9.3	Aggregate output, inputs and productivity	176

List of charts and graphs

5.1-5.7	Comparative machine parameters	109–12
7.1	The structure of co-operation and specialization institutions in Comecon	149

Introduction

Since the late 1950s there has been a general slowdown in the rate of Soviet economic growth. The phenomenon is now well known to Western specialists on the Soviet economy. Its broader implications, with regard to the difficulty of allocating resources between competing policy objectives and the political tensions which arise from this process, have been analysed by Seweryn Bialer, Abraham Becker, Philip Hanson and many others. At the heart of the problem is the failure of the central planning mechanism, which took shape in the 1930s under Stalin's political direction, to promote rapid technical progress. Technological development has, of course, taken place in the USSR during the past two decades, but not at a sufficiently high rate or sufficiently broadly to override the effects of growing resource scarcity; to express the problem in Soviet parlance – a successful transition has not yet been made from the stage of '*extensive* development' to one of '*intensive* development'. In most industries a substantial technology gap still separates the Soviet Union from the advanced Western countries. Mr Gorbachev himself has characterized this as 'problem number one' for the USSR, and in a key-note address to a special Central Committee Conference in June 1985 reaffirmed that:

The Party views the acceleration of scientific and technical progress as the main direction of its economic strategy, as the main lever for the intensification of the national economy and for raising its efficiency; and hence, for the solution of all other economic and social issues. These tasks are so pressing that action has to be taken without losing any time.

Although it would be quite wrong to view the present Soviet difficulties in crisis terms, one nevertheless gets a strong sense that the country under its new leader is moving towards a series of crucial decisions, which will determine the shape of Soviet development for years to come. For those of us with a special interest in Soviet science and technology, therefore, it seemed an appropriate moment to take stock of the situation and to establish a basis for interpreting the significant policy objectives and organizational changes which are likely to be announced at the next Party Congress in early 1986.

In the Autumn of 1984, some months before Gorbachev took over formally as General Secretary of the Party, a Symposium on Soviet Science and Technology was held at Birmingham University under the joint auspices of the

Centre for Russian and East European Studies (CREES) and the Department of Extramural Studies. The symposium was unusual in that half the participants were academics and the other half were professional analysts drawn from the British Government and defence community. The intention which lay behind the symposium was to bring the academics into closer discussion with fellow specialists, who shared common concerns and who were both consumers and critics of our published work. It was clearly understood at the outset that those participating on the government side would be doing so as individuals and not as official representatives and, moreover, that the subject matter for discussion would be confined entirely to what was available in openly published sources. The various chapters in this book are revised and updated versions of the papers presented at the symposium.

The chapters fall into one of two major but related themes. First, there is a group of chapters concerned in various ways with Soviet technological *performance*, a longstanding interest of researchers at CREES. Amann's introductory chapter reviews recent evidence about Soviet technical progress on the basis of a wide range of criteria, and poses some of the general issues which are taken up in later chapters. Cooper presents some detailed results arising from his recent work on the Soviet defence sector, hitherto considered to be an exception to the general run of civilian research and development and to some extent pursued at its expense. Hill and McKay evaluate Soviet performance in two traditional sectors (machine tools and electric motors) while Snell and Rimmington extend the range of existing case-studies into the crucial spheres of electronics and biotechnology where very little work, using Soviet sources, has been done to date. A second group of chapters attempts to evaluate the various options which the Soviet Government might explore either singly or jointly, in order to *improve current performance* and thus to accelerate the tempo of scientific and technical progress. Of the possible external solutions, Bertsch looks at the present and future prospects for acquiring advanced Western technologies, while Sobell examines the extent to which co-operation within the Soviet bloc itself might help to stimulate more rapid technological development. From an internal perspective, Dyker analyses the likelihood, given past experience, that institutional reform will have any significant or durable impact. Finally, Bond puts the whole discussion in its broader context and advances a number of bold forecasts of Soviet economic performance to the end of the present decade.

Although the chapters in the volume have a thematic coherence, the reader will soon discover that the general line of interpretation is not consistent throughout. There was widespread disagreement throughout the discussions which took place at the symposium, and this indeed, was one of its most valuable and enjoyable aspects. On the whole, and broadly for the same reasons (endemic bureaucratic rigidity – short of a truly radical reform – which affects both domestic technology and the diffusion of foreign imports), Amann, Dyker, Sobell and Bertsch are inclined to be relatively pessimistic about Soviet prospects, though all would accept that modest improvements could take place as a result of incremental reform, combined with a greater sense of political purpose. However, none of these changes would be likely to transform Soviet technological performance in a fundamental way. This view is consistent with the previous work of the Birmingham group and is widespread among Western specialists. However, other chapters in the book advance, in varying degrees, strongly revisionist views, which challenge previous thinking on the subject. Instead of the Soviet defence sector existing in a 'world of its own' and acting as a drag on the civilian economy, Cooper emphasizes the strong links which exist between the two sectors, regarding the former as a potential dynamo of good industrial practice. Snell details the impressive list of microprocessors which are described in current Soviet literature and observes that even the undoubted copying of Western designs presupposes an advanced technological capability; moreover, bit-slicing is seen as a viable substitute or supplement to large scale integration (LSI). Rimmington describes the development and manufacture of single cell protein within the framework of what is probably the world's largest microbiological industry, and explains the overall policy context. Hill and McKay, by means of a detailed analysis of state standards, demonstrate that some basic traditional technologies are on a par with those in the West: electric motors exhibit comparable performance over a range of criteria; the initial alignment accuracies of Soviet general-purpose machine tools are also comparable to those in the West, though they are prone to lower levels of reliability and durability in use. The debate here is not simply an inevitable difference in perspective between the generalists and the particularists (Bond is firmly in the camp of the optimists) but centres on a clash of conceptions and judgements which have not yet been definitively resolved by empirical evidence. Detailed monitoring and analysis of Soviet policies and performance over the next five years will tell us a good deal more and may clarify some of these issues. It will be a crucial time for the USSR.

The editors and authors would like to thank all the participants at the Symposium and the following individuals, in particular, for their comments and criticisms: Mr Peter Smith (Ministry of Defence), Wing Commander Mike Grigson (RAF), Mr Graham Kidd (ICI Agricultural Division), Dr Daniel Franklin (*The Economist*), Dr Christopher Davis (CREES, University of Birmingham), Dr Martin Cave (Department of Economics, Brunel University) and Mr Jonathan Stern (Joint Energy Programme, Royal Institute of International Affairs). Any faults are, of course, the responsibility of the authors themselves. But they would certainly have been more numerous without the informed comment from which we all benefited. We would also like to thank the Royal Air Force for supporting the Symposium on the services side: particularly Air Commodore Tony Mason and Group Captain Tim Garden (Director of Defence Studies, RAF) for their generous help and encouragement of a novel event, which could not easily be handled through the normal channels. Mrs Jane Hamilton-Eddy (Ministry of Defence) was equally helpful in coordinating the participation of government research staff. A degree of administrative flexibility was demonstrated – not a notable feature of the bureaucratic machine we were studying! The Department of Extramural Studies provided first-class organizational support for the Symposium, and special thanks are due to its Director, Mr R. Sawers and to Mrs Ann Hollows, who made many of the detailed arrangements. Individual authors acknowledge the help they have received from various research foundations in their own chapters but it is appropriate here to express a general acknowledgement to the Economic and Social Research Council (ESRC) for its consistently generous support of the work on Soviet science and technology carried out at CREES. Finally, we would like to express our sincere thanks to Sandra Cumberland and her colleagues Lesley Woolley and Julie Cant in the CREES office for the very efficient way in which they have helped us to prepare this material for publication. In this matter, as in others, power rests ultimately in the hands of the secretariat.

Ronald Amann
Julian Cooper
June 1985

Postscript

Since the book went to press a number of key officials mentioned in the book have been replaced, including Tikhonov, Chairman of the USSR Council of Ministers, by N. I. Ryzhkov; Smirnov, Chairman of the Military-Industrial Commission, by Yu. D. Maslyukov; Dmitriev, Head of Defence Industry Department of the Central Committee, by O. S. Belyakov; Shokin, Minister of the Electronics Industry, by V. G. Kolesnikov; and Rychkov, Head of the Main Administration of the Microbiological Industry, by V. A. Bykov.

1
Technical Progress and Soviet Economic Development: Setting the Scene

RONALD AMANN

In a book published in 1977, which was based on data extending to the mid-1970s, my colleagues and I reached the conclusion that, with the notable exceptions of some priority sectors, Soviet technological performance was inferior to that of the advanced Western industrialized countries and, more controversially, that the USSR had demonstrated no strong signs of catching up during the previous 15–20 years.[1] Progress had certainly been made, but in most of the sectors studied the technology gap had not been closed; instead, the USSR found itself chasing a rapidly moving target. In 1982, with the benefit of data extending to the late 1970s, this broad judgement was confirmed in a second book and an attempt was made to explain the reasons for this unsatisfactory performance.[2]

Much has happened since these two books were published. Technological inertia has come to be seen, both among Western analysts and in the USSR itself, as perhaps *the* crucial constraint on future Soviet economic development. This has given rise to some interesting Western writings and, on the Soviet side, to a number of unusually frank and penetrating analyses, which began to appear in the press and economic journals towards the end of the Brezhnev era. One could discern in these latter writings a degree of frustration and impatience: a feeling that for too long the process of economic and political decision-making had been locked on 'automatic pilot' and that the time had come for the restoration of manual controls by a more vigorous leadership. The purpose of this introductory chapter is to assess some of the recent evidence with particular reference to (1) technological performance; (2) the reasons for it; and (3) the feasible options for improving it. These general themes are pursued in greater depth throughout the remaining chapters of the book.

The pessimism of the 'Birmingham group' did not escape criticism from academic colleagues in the West, especially during the first stages of our work. While accepting the usefulness of a disaggregated case-study approach, some readers were put off by the artificiality of calculating technological leads and lags in precise numbers of years. To a large extent we would agree with this criticism. The sectoral assessments of relative technological level are more approximate than they might seem, but, on the other hand, there is no alternative to the systematic assembly of data on diffusion of new technologies and comparative dates of first commercial production. We were keenly aware that in the aggregate this could yield only a general indication of performance, no matter how scrupulous we tried to be in the selection of the sample. Other readers, focusing on Soviet achievements in space exploration, on the technological basis of Russia's formidable military power and on the enormous scale of the Soviet support for research and development in both expenditure and manpower terms, found it hard to accept that the overall picture could be quite as black as we had painted it. Perhaps the most sophisticated objection to our conclusions was that they appeared to be out of line with a well established international pattern of technological development whereby backward nations, taking advantage of their ability to borrow cheaply, would inevitably catch up with the most advanced countries in the long term.[3] Leaving aside what one means by 'the long term', this argument probably underplays both the specific institutional features of the Soviet system, adversely influencing its capacity to absorb foreign technologies, and the impact of a quantum leap between major phases of technological development, which could reopen gaps between leaders and followers. Arguably, it is precisely such a prospect which greatly concerns Soviet leaders at the present time as they observe the rapid development of electronics and biotechnology in the West and weigh its political and economic consequences for themselves.

Despite the reservations of the sceptics, some of which we would accept ourselves, it would seem that in the light of economic trends in the late 1970s and in the first half of the 1980s the conclusions of the Birmingham group have been broadly right. Although we did not explicitly predict the marked slowdown in the rate of economic growth during this period (much of which is the result of non-systemic factors) the phenomenon is consistent with our analysis and conclusions. Moreover, our central notion that there exist powerful institutional impediments to future technical progress, which can not be easily overcome because of *deep historical influences*[4] on popular attitudes and organizational behaviour, anticipated the main elements of the internal reform debate which got underway during the early Andropov period.[5] In particular, if one ignores the Marxist terminology, the parallels between our conclusions and the position advanced so forcefully by Tatyana Zaslavskaya are quite striking.[6]

Given our general view that the behavioural characteristics of the Soviet economic system are rooted in history, I would not expect that the creation of new planning bodies or superficial amalgamations designed to strengthen the link between science and production or the rearrangement of the incentive structure would necessarily lead to a sustained acceleration of technical progress. Incremental reforms of this kind would encounter a force-field of resistance from officials at all levels of the system, who had assimilated the established rules of the game and had perfected various survival skills. Thus, if we had been forced to predict the likely pattern of Soviet technological development during the first half of the 1980s on the basis of our previous knowledge of institutional performance, the outlook would not have seemed especially bright. What are the facts now available?

TECHNOLOGICAL PERFORMANCE

There is a spectrum of indicators of technical progress ranging from aggregated but rather indirect ones at one end to more direct but potentially unrepresentative ones at the other. No one indicator is sufficient in itself, and in attempting a general assessment of recent Soviet technological performance one is looking for a substantial degree of consistency between these various measures.

Aggregated economic indicators

The most general and aggregated indicator of all is the annual average rate of economic growth. According to Western estimates Soviet GNP grew on average by 6–7 per cent in the 1950s, 5 per cent in the 1960s and 3 per cent in the 1970s. It is unlikely to rise much above 2 per cent in the 1980s.[7] According to the latest plan fulfilment results available at the time of writing, the slight improvement which took place in 1983, perhaps as a result of Andropov's campaign to tighten up on work discipline and the fulfilment of contract deliveries, began to lose momentum once again in 1984.[8] Of course there are many climatic and environmental factors which depress growth rates but it is interesting to note that one of the most outspoken observers of the Soviet scene, Academician V. Trapeznikov, has discounted the influence of weather and the depletion of sources of energy and raw materials in the most accessible areas of the country and has laid the blame squarely on an inadequate rate of technical progress – and particularly on the failure of unimaginative central planners to appreciate its true importance.[9]

If overall economic growth is too gross an indicator to provide a satisfactory perspective on the pace of technological development, it follows that a better approach would be to focus on the effectiveness by which resources are used: on the various measures of productivity, which form an integral part of the Soviet objective of 'intensive development' towards which they are constantly striving. As the official Soviet figures show (see table 1A.5) there has been a marked decline in the annual rate of growth of labour productivity during the last decade or so. It is currently running at about 2.5 per cent, less than half the going rate in the early 1970s and in some industrial sectors such as ferrous metallurgy the decline has been particularly steep. Given the declared objective of the planners that future economic growth must rest exclusively upon greater productivity, this trend must be distinctly worrying.

Trends in labour productivity certainly do reflect the rate of technical change in the economy, embodied in new capital and in the growing skills (know-how) of the labour force, but they are also influenced by other factors such as the size of the capital stock, its intensity of utilization and the motivation of workers. These are not, strictly speaking, part of 'technical progress'. Similar drawbacks also apply to the use of capital productivity as an indicator of technical progress. Here, the size and qualifications of the work-force, the utilization of plant and the infrastructural costs associated with the procurement of fuels and raw materials are all important determinants of capital productivity, which do not necessarily depend upon rapid technical progress. On the other hand, technical progress would clearly have a major influence on this indicator if, as often happens, new Soviet equipment did not meet the best world standards. The fact that the annual retirement rates of obsolete plant and equipment in the USSR are on average about half those in the USA,[10] while the slow introduction and assimilation of new capital stock as a result of design failures and construction delays often leads to 'moral obsolescence', would also have a potent impact. In fact, during the last 20 years capital-output ratios in the Soviet Union have more than doubled, signifying a marked decline in capital productivity.[11] Because of the slow growth of the labour force, the introduction of new automated capital equipment has often simply meant that new jobs have been created which have remained unfilled. The substitution of capital for labour as a source of productivity has thus proved difficult.[12]

Since the growth of labour productivity and capital productivity at any given point in time depends upon the relative proportions of labour and capital, a more satisfactory measure of technical progress is one which analyses total factor productivity. According to US government analysts, the rate of growth of total factor productivity in the USSR after 1973 declined at an average rate of 0.8 per cent per annum.[13] A recent CIA assessment shows that the annual average growth rate of total factor productivity of GNP picked up slightly in the last few years, but is still either negative or negligible.[14] Of course, crude estimates of total factor productivity and its growth over time once again contain a large number of non-technological influences (negative influences such as natural resource exhaustion and bad weather, and positive ones such as improvements in labour quality due to educational advance, changes in the balance of employment between agriculture and industry, economies of scale and effects of planning reforms). But in principle some or all of these influences can be allowed for in the calculation, leaving a residual which approximates to 'technical progress'. No doubt these adjustments are subject to error and must remain controversial but they are well worth attempting. We have summarized some of the major works in this field in one of our previous books.[15] In a more recent attempt to measure what he calls 'technical progress proper' (TPP) the distinguished American economist Abram Bergson concludes that by the mid-1970s the annual rate of growth of TPP in the USSR had fallen to 0.16 per cent; this performance in Bergson's opinion was 'within the range of Western experience but inferior to that of Western countries at a comparable stage of development.'[16]

Indicators of the recent slow rate of technical progress in the USSR, contained in these aggregate economic measures, receive further confirmation when we examine the pattern of Soviet foreign trade. It is a characteristic of highly advanced countries that value-added manufactures form a substantial proportion of their total exports; exports of machinery and equipment are particularly prominent. At the present time, however, only 6 per cent of Soviet hard currency exports are accounted for by sales of machinery and equipment, a proportion which has remained more or less unchanged since 1970. Approximately two-thirds of Soviet hard currency earnings are derived from energy exports, oil in particular accounting for about one half of the total. By contrast, machinery and equipment currently account for nearly one quarter of hard currency *imports*, a slight decline since the mid-1970s when the impact and political significance of the East European debt began to be felt.[17] Even more worrying from the Soviet and East European view is the emergence of newly developed countries as serious exporters of manufactured goods to the established industrial countries of the Organisation for Economic Cooperation and Development (OECD). The share of Eastern Europe in OECD imports of machinery and equipment, for example, increased from 0.1 to 0.7 per cent during the 1970s, while the share of only six newly industrialized countries (NICs) rose from 0.1 to 4.1 per cent during the same period.[18] The fear of some Western governments during the detente era that exports of advanced plant and equipment to the USSR and the other countries of the Council for Mutual Economic Assistance (CMEA) would eventually 'boomerang' back in the form of increased manufactured exports on Western markets have proved groundless up to the present time. Instead, the USSR now faces the challenge of raising the quality and novelty of its products in order to maintain its position.

More direct general indicators of technical progress

The annual Soviet statistical handbook, *Narodnoe Khozyaistvo SSSR*, contains a series of statistics which relate directly to the rate of technical progress. They have been ignored previously by Western researchers mainly because the statistical categories are somewhat vague, it is difficult to relate them to any comparable Western figures and thus there is a shrewd suspicion that the figures could be inflated arbitrarily in order to make Soviet performance look as good as possible. On the other hand, as Vladimir Kontorovich has pointed out,[19] the long-term trends could still be interesting, perhaps *especially* because of the upward bias one would expect to encounter.

The Soviet figures (presented in tables 1A.1–4 (see pages 26–30)), cover the whole of the latter part of the research–production cycle: invention, innovation, diffusion and incremental improvement. Compared with achievements in scientific research, these are the phases of development where it is generally thought that the USSR exhibits relative weakness as a result of discontinuities between different phases of development and lack of incentives. Ideally, the figures in value terms should be systematically adjusted for inflation but without a suitable price index this is very difficult to do; in any case, the main point in looking at these statistics is to discern the relative balance between the growth of innovation activity and its economic impact (both of which would have to be deflated). The general picture that emerges from the statistics is extraordinarily striking and consistent. There has been a continuous absolute decline since the early 1960s in the creation of prototypes of new machines and equipment, especially in the more advanced science-based industries (table 1A.1). Though growth rates in terms of physical units will have been restrained to some extent by the 'sophistication factor' (the tendency towards higher development costs per prototype at higher levels of technological advancement) it is unlikely that this factor can account entirely for this remarkable trend. Table 1A.2 deals with the next stage in the research–production cycle: the introduction of new technologies into the economy and their economic impact. Once again one can see a fall in the rate of growth of these measures since the first half of the 1970s, even in value terms (though the relatively more rapid decline of the physical indicator here gives some credence to the unit cost argument). The most startling trend to emerge from table 1A.2, however, is the relatively steep decline of the impact indicators, expressed in terms of labour savings from new technologies and their annual economic effect. This suggests a growing ineffectiveness of industrial innovation. If it were not for the slight peak in performance in 1983 the annual economic effect from the introduction of new technologies would have come to rest at zero growth; presumably, if one allowed for the exaggerated estimates of economic effect, which are an endemic feature of the Soviet R and D system, the position would appear even worse.

Though the distinction between the coverage of tables 1A.2 and 1A.3 is not absolutely clear, for practical purposes we can treat the former as relating to the first-time introduction of new technologies, while the latter deals with their subsequent diffusion throughout the economy. It would seem, therefore, that despite a slight acceleration in the rate of diffusion in the latter half of the 1970s the general trend over the last 15 years has been downwards. In the crucial technological areas of machine-building and chemicals, growth rates have become negative (the high 1971–5 figure for chemicals is almost certainly influenced by substantial imports of plant and equipment from the West). It is possible that these figures could give a false impression if the Soviet planners had succeeded in concentrating resources on fewer objectives and thus in increasing the unit-value of industrial modernization schemes. However, there is no evidence that the phenomenon of *raspylenie sredstv* has been eliminated and, in any case, one would have expected that the current change in emphasis from new construction on green-field sites to modernization of existing plants would have tended to increase the number of units.

Table 1A.4 focuses mainly on the incremental improvements at the shop floor level, which are such an important aspect of technical progress in the West. This is where science interacts with production. In the course of fully mastering and modifying a new process fundamentally new technologies can begin to take shape. In the USSR, however, the long-term statistical trends would seem to cast some doubt on the vitality of these creative responses. Since the early 1970s there has been a fall in the growth rate of the number of improvements introduced, and a similar though less pronounced fall in expenditures on their implementation. More particularly, there was an abrupt decline in the growth of the economic effect of these measures at the end of the 1970s, corresponding to the general slackness and inertia which was characteristic of the final years of the Brezhnev era.

We might reasonably conclude from tables 1A.1–4, which almost certainly have an upward bias, that even if Soviet leaders were unsophisticated enough to believe their own official figures they would still have grounds for pessimism.

Disaggregated data and sectoral studies

An alternative approach to the measurement of technical progress in the aggregate is to break down the above general categories into their sectoral components. In a recent paper, Julian Cooper[20] has updated and extended somewhat the figures which we assembled in our 1977 book on innovation (comparative dates of first commercial production) and the subsequent diffusion of new technologies. It is possible to summarize only the highlights of that analysis here and to take up some of the general issues of interpretation which he raises.

Cooper's updated figures on first commercial production indicate a respectable performance by the Soviet Union in some traditional or high priority industries (iron and steel and nuclear power, for example), but substantial lags in other traditional industries and in the whole civilian electronics-related sector: computers, industrial robots, sophisticated consumer durables and, of course, electronic components themselves. These indicators of comparative lag in the initial application of new ideas into industrial practice are consistent with the findings which emerge from Martens and Young's novel study of Soviet patents.[21] The authors concluded that two years after Soviet inventions had received an author's certificate, the equivalent of a patent in Western countries, only 23 per cent of them had been implemented in production; the comparable figure for the USA was 66 per cent and for West Germany 64 per cent. The possibility that relative delays in implementing Soviet 'patents' might owe more to their inherent lack of quality than to the sluggishness of the economic system would, of course, be of no comfort whatever to the Soviet authorities.

It is the process of diffusing of new technologies rather than their first commercial production, however, which is the major determinant of technical progress. A nation can overcome the fact that it has not played a major role in pioneering new industrial technologies. Provided that it can incorporate them rapidly, on a mass scale, as Japan succeeded in doing during the post-war period, it can remain competitive and eventually come to assume a leading role at the technological frontier. Cooper's data, which analyse the proportion of new technologies in general product groups and the rate of growth of those proportions over time, show that particularly good progress has been made during the last decade in the Soviet pulp and paper industry (the diffusion of the sulphate process), in the manufacture of colour television sets and in the fertilizer industry[22] (with considerable Western technical assistance). However, there are clearly serious lags across the whole data-processing and electronics sector. In 1978 the USSR had only 84 mainframe computers per million population compared with 898 in the USA and 392 in Japan. A Soviet personal computer, the Agat, is available in only strictly limited quantities at present and according to a Western specialist who has tested it extensively it is derivative in design yet not competitive with Western personal computers in terms of either price or quality.[23] In fact one may question how prepared the USSR is for the rapid horizontal communications of the electronic age when the number of

Table 1.1 Adoption of new technologies: dates of first commercial production

Technology	USSR	USA	Japan	FRG	UK
Oxygen steel	1956	1954	1957	1955	1960
Continuously-cast steel	1955	1962	1960	1954	1958
Synthetic fibre (nylon)	1948	1938	1942	1941	1941
High-pressure polythene	(1953)	1941	1954	1944	1937
Nuclear power station	1954	1957	1966	1961	1956
Numerically controlled machine tools	1965	1957	1964	1963	1966

Source: Compiled from R. Amann, J. M. Cooper and R. W. Davies (eds), *The Technological Level of Soviet Industry* (1977).

telephones per 1,000 of population is equivalent to that of the USA in the late 1930s and even proper telephone directories are not widely available.

The nub of the problem from the Soviet point of view can be seen in the relationship between the data in table 1.1 and table 1.2. In table 1.1 we have a selection of technologies which, with the exception of the chemical branches, the USSR was reasonably quick to introduce initially on an industrial scale. This is especially true of the metallurgical technologies. The Soviet Union was among the first countries in the world to introduce oxygen steel and continuous casting into full-scale production. Yet when we look at table 1.2 it is evident that in the USSR these technologies have not spread as rapidly throughout their particular industrial branches as they have done in Western countries. The Soviet pattern is out of line with the others in each case, including even synthetic fibres and polymerized plastics, where massive quantities of advanced plant and equipment were acquired from the West during the 1960s and 1970s. Evidence also suggests that the newly industrializing countries, admittedly from a much smaller industrial base, have incorporated these technologies more rapidly than the USSR and have thus begun to gain a competitive edge in terms of their level of technological modernity and overall economic effectiveness. To take only one example, the proportion of oxygen steel in total steel output in the USSR increased from 3 per cent to 29 per cent between 1960–80; in Brazil during the same period the proportion increased to 63.4 per cent and in South Korea to 69.7 per cent.[24] In part, Soviet performance is held back by the general slowness of creating new production facilities but also, more particularly, by the failure of new technologies to drive out or 'extinguish' obsolete ones.[25] Whatever the principal reason for this phenomenon, whether it is conservatism pure and simple or a conscious desire to retain old facilities as a buffer against scarcity of commodities at relatively low levels of national per capita output, it acts as a brake on the pace of technical progress.

When we contemplate the above statistical trends it is worth observing that in general they probably have an inherent tendency to *flatter* Soviet performance.

Table 1.2 Subsequent diffusion of new technologies: patterns of output within commodity categories in 1982 (%)

Technology	USSR	USA	Japan	FRG	UK
Oxygen steel (as % of total steel)	29.6	62.1	73.4	80.9	66.1
Continuously cast steel (as % total steel)	12.1	27.6	78.7	61.9	38.9
Synthetic fibres (as % total man-made fibres)	51.2	91.2	83.8	83.1	78.6
Polymerized plastics (as % total plastics)	46.4	87.5	80.0	73.0	79.3
Energy generated by nuclear power stations (% total)	7.1	12.4	17.6	17.3	16.7
NC machine tools (as % total value of metal-cutting machine tools[a]	16.6	34.0	52.8	20.6	27.7

[a] 1980.

Sources: Julian Cooper, 'Is there a technological gap between East and West?', paper prepared for conference organized by Canadian Institute of International Affairs, Toronto, June 1984; *UN Yearbook of Industrial Statistics*, Vol. II, 1983, pp. 375–80 and pp. 382–4; *Statisticheskii ezhegodnik stran chlenov SEV*, 1983, pp. 95–6.

The most advanced technologies at any given point in time, by their very nature, have not become established enough to be the subject of statistical recording. To take a specific example, one of the most significant recent trends in the chemical industry in the West is product diversification through the manufacture of small-batch, specialized, high-quality plastics and catalysts, rather than the continued expansion of bulk polymers.[26] This is an area where the USSR is notoriously weak,[27] but that weakness does not yet find expression in precise statistical form.

Equally, one should make the obvious point that the *general* indications of technological backwardness revealed by the above statistics do not necessarily mean that *all* sectors of Soviet technology are backward. The more that one looks in detail at Soviet performance by means of specific case-studies, the more variations in technological performance begin to come into focus. This was one of the main reasons why we adopted the case-study approach in our previous work. Since then, a number of other interesting studies and assessments have become available of technologies which developed rapidly during the 1980s: for example, the whole computer and electronics area,[28] defence technology,[29] the gas industry,[30] and industrial robots.[31] Two further case-studies by Paul Snell and Tony Rimmington on the key areas of microprocessors and biotechnology, based on a thorough survey of Soviet sources, are included later in this book. An evaluation of this latest batch of studies is beyond the scope of this introductory chapter. It is sufficient to say that a careful reading of them will effectively dispel any oversimplified notions that a general tendency towards technological inertia necessarily adds up to abject failure across the board. It clearly does not. Soviet society, like any other, has its dynamic individuals, who can mount effective 'campaigns' for technological development, and the economic system is capable of concentrating resources on key technological objectives once these are defined and understood. As John Kiser (among others) has shown,[32] one can without too much difficulty find several examples of Soviet success in selling technology to Western countries in those specific areas where they have succeeded in developing and implementing new ideas. During the 1970s, US companies acquired 126 licences from the USSR and Eastern Europe.[33] But these examples, which are a sobering riposte to those who believe that East–West technology transfer is a 'one way street', are the exception rather than the rule.

Interpretations and counter-interpretations

Although the general thrust of the above evidence seems to point in a definite direction there are, in fact, different ways of looking at it and therefore considerable scope for continued argument. Hans Hohmann[34] has pointed out to me that although the trend of the published statistics undeniably suggests that the USSR is not succeeding in 'catching up and overtaking the advanced Western countries' (as Stalin once put it) the fact that they are managing to keep up, albeit lagging somewhat, represents a considerable amount of technical progress. Provided that the Soviet Union can maintain this position without falling further behind, importing technology relatively cheaply in selected areas, it would be an exaggeration to represent the present level of performance as a crisis. A corollary to this argument is that incremental reform rather than a fundamental transformation of the planning system would probably be sufficient to keep the economic engine ticking over at an acceptable speed. While I would accept the first part of Hohmann's argument I would take issue with the second part. Clearly there *has* been substantial technical progress in the USSR during the last decade or so, and the economy is by no means on the verge of collapse. On the other hand, the very fact that Soviet leaders themselves have recently expressed such clear anxieties about the level of technology in their country suggests that the rate of progress is *not* acceptable. These worries relate to the general deterioration of economic indicators described above, the internal political implications of scarcity, the loss of international prestige and influence as a result of visible economic failings, the inability to guarantee the future purchase of advanced Western equipment at a time of revolutionary developments in technology and a hostile US administration and, more generally, the psychological impact of seeing one's central ideological expectations of future progress so blatantly contradicted. I have explored this relationship between technical progress and political change in the USSR in more detail elsewhere.[35] What all this means in terms of future policy responses is, of course, a matter of judgement and no doubt will continue to be debated between Dr Hohmann, myself and many others. Some of the possible options are summarized in the final section of this chapter.

An alternative line of interpretation, with rather similar policy implications to those inherent in Hohmann's position, has been advanced by my colleague Julian Cooper.[36] Cooper's main contention is that visible indicators of technological performance give only a very partial indication of the reality which lurks beneath the surface. His first major point is that statistics of diffusion can not tell us much about Soviet *capability* in terms of know-how, especially when the considerable scale of the Soviet R and D effort is taken into account. Superficial Western assessments thus seriously underestimate the future potential of the Soviet economic system to generate technical progress. However, one might object that the rapid diffusion of new technologies, as a result of 'learning by doing' is in some sense a *precondition* for future technological advance: a much more potent one than the abstract potential of Soviet research institutes with their endemic ivory tower attitudes. Also, if this alleged 'capability' has existed over a long period it is not clear why, apart from selective instances, it has not become reflected in visible and measurable performance. If Cooper is saying that Soviet achievements in scientific research are much more impressive than those in the sphere of industrial innovation, he is repeating a point which is already recognized in most Western writing on the subject. The argument therefore resolves itself to a matter of semantics and gut optimism. What Cooper regards as 'capability' others, including myself, might regard as an indication of systemic weakness, though we would both agree that the system has the capacity to promote the selective development of priority areas.

A second line of argument is that the USSR is in many ways not necessarily technologically *backward* but *different*. One must constantly relate the pattern of a country's technological development to its specific circumstances in order to determine whether a given pattern is appropriate or not. To proceed otherwise is to apply mistaken criteria and to introduce ethnocentric biases – especially the blind acceptance of a general trend towards technological complexity. A paradigm case here is that of automatic washing machines, which are not widely diffused among Soviet households because electrical circuits in Soviet apartments have insufficient capacity to support them. Simpler manual machines are much more 'appropriate', though automatic machines can and have been developed (in the circumstances, the absurdity of doing this seems remarkable!). A more serious example might be the trade-off between technological sophistication and rugged combat effectiveness in Soviet military technology. In general terms this is a formidable argument and one which is difficult to refute. To be truly persuasive however, Cooper would need to demonstrate that the USSR was ahead of the West in an appreciable number of technological sectors. If not, it is difficult to accept that what we see and measure is not a gap, ingeniously minimized by improvisation, but a different set of priorities. Where have these alternative priorities been explicitly announced and where have they reached successful fruition in comparative terms? A further reservation is that if a nation is compelled by its poorly developed infrastructure to opt consistently for relatively simple ('appropriate') technologies, is not this in itself a measure of technological backwardness?

It would be interesting to pursue some of these issues in future work. However, though one recognizes the possibility of alternative interpretations the most obvious one still seems to be the most plausible. Taking into account the whole spectrum of indicators of technical progress, a consistent picture of inertia emerges which must be (and, in fact, is) a matter of serious concern to the Soviet political leadership. Moreover, as we shall see below, the present way in which the R and D system functions gives no grounds for anticipating substantial improvement in the short term, and the alternative strategies for bringing about a decisive improvement are likely to be either politically adventurous or limited in their effectiveness.

SYSTEMIC INFLUENCES ON INDUSTRIAL INNOVATION

During the last 30 years, as the economic framework created by Stalin in the 1930s encountered the new economic and technological demands of the postwar period, the promotion of industrial innovation has become a major preoccupation in the USSR. Despite countless attempts to reform the organization of the R and D system and to increase bonuses and other kinds of economic incentives, three major themes have continued to occupy a central place in scholarly discussion:

(1) The economic system is permeated with departmental barriers, which impede co-operation between different industrial ministries and tend to isolate scientific activity from manufacturing activity. Because of informational overload the central planners can not act as an effective counterweight to this fragmentation.
(2) There is a systemic tendency on the part of industrial enterprises and their ministries to maximize output at the expense of the quality and novelty of products and, thus, to attach a relatively low priority to industrial innovation.
(3) The inadequacy of pilot plant and large-scale testing facilities creates an imbalance between resources devoted to research and development. This acts as a bottleneck, holding back the flow of ideas and their conversion into designs of new products or processes.

Successful innovations occur but they invariably do so against the prevailing tide as a result of political will, individual initiative or, more often, both. It is not by accident that the successful innovator is typically portrayed as a 'hero' in Soviet industrial novels (see, for example, Dudintsev's *Not By Bread Alone*). He frequently needs to be heroic when confronting the forces of bureaucracy! These basic features of the Soviet system are well understood both in the USSR and in the West and do not require further elaboration here – nevertheless it is interesting to look at the new developments which have taken place during the last five years, and which could modify our picture of general systemic characteristics.

First, it is apparent that departmental barriers continue to hamper the innovative process and have become such a serious matter of concern that a general reform of the whole ministerial structure has been proposed by some Soviet writers.[37] Though not going this far, V. Kudinov, Deputy Chairman of the State Committee for Science and Technology (SCST), has pointed out that deployment of R and D resources according to the branch principle runs counter to the emerging inter-branch character of technical progress. As a result, some important projects 'fall into a kind of no man's land'.[38] Central co-ordination and control are essential prerequisites for implementing the 170 complex projects, which comprise the national programme for scientific and technological development under the formal authority of SCST. But, in reality, the State Committee depends on the goodwill of the ministries to do its bidding; it is not a powerful 'boss' able to command respect.[39] As a result, ministries often fail to provide the resources laid down as their contribution to national programmes, and their head institutes continue to defy requests to modernize the branches' product range, their monopoly position unrestrained.[40] Even the scientific councils formed to co-ordinate work on each complex programme appear to have been captured by departmental interests.[41] G. Marchuk, the Chairman of SCST has observed that:

Ministries and departments do not provide the ear-marked financial and material resources for the tasks laid down in the programmes, especially at the stages of introducing and assimilating new technology. This in its turn means that under conditions where labour, material-technical resources and capital investment are in short supply, branch ministries give preference to the manufacture of already assimilated products, not infrequently relegating new technology to a secondary order of importance.[42]

One consequence of the inability of the State Committee for Science and Technology to impose a unified national perspective on the development of industrial R and D has been the growing authority of the USSR Academy of Sciences in the sphere of applied research. In the early 1960s, as the result of a stormy internal debate, it seemed as if the Academy would henceforth confine itself to initiating and co-ordinating the main directions of fundamental research throughout the country. Its Division of Technical Sciences was disbanded. However, confronted with unsatisfactory performance on the industrial front it was not long before the Soviet Government once again had to draw upon the scientific authority and expertise of the Academy – just as it did during the industrialization drive of the 1930s. The Academy is a reliable centre of scholarly excellence and its scientists have increasingly come to be used as shock troops for overcoming critical areas of technological backwardness.[43] Based on the original model of the Siberian Division of the Academy in Novosibirsk, there has been a substantial growth of regional scientific centres (especially in the Ukraine), which attempt to integrate science and production, and which are reinforced by strong party support. The fact that Academy institutes are assuming greater responsibility for practical results provokes, of course, an inevitable demand on their part for greater access to development facilities, which would assist in the achievement of these results. The trend toward practical involvement therefore deepens. A series of new research institutes in the microelectronics area has been created under the Academy with the resources, at least in theory, to develop new computer technologies through all stages of preparation up to and including their introduction into the economy. A new Division of Information Sciences, Computer Technology and Automation has been formed to co-ordinate these activities.[44] Even more far-reaching suggestions include the creation of a new international centre to promote work on robotics (equivalent to the role played by the Dubna centre in nuclear research within the Eastern bloc); it has even been suggested that a chain of science-production associations, based on new scientific discoveries, should be set up within the Academy.[45] These developments have enhanced the Academy's political stature. From indications at recent Party Congresses and in the deferential tones with which leading officials of the SCST now refer to it, it seems clear that the USSR Academy of Sciences is emerging as a senior partner in the national co-ordination of Soviet research and development. This impression is reinforced by Gorbachev's speech to the special Central Committee Conference on technical progress in June 1985.

The organizational framework and its unintentional encouragement of fragmentation and isolation is, of course, only one aspect of the innovation process. Another major aspect concerns more directly the motivations of the individuals and institutions working within this framework who are influenced powerfully by the prevailing structure of economic incentives. Bonuses for successful R and D and innovation are a longstanding feature of the Soviet economic planning system, but since the late 1970s, a concerted attempt has been made to increase their impact in comparison with the much more substantial bonuses that industrial enterprises could traditionally earn for successful fulfilment of the output plan. A robust assault was therefore to be made on the primacy of gross industrial output ('*val*'). From 1979 onwards there was a rapid growth of price increments permitted on the basis of estimates of the economic return of new products. Enterprises and R and D organizations were also given the right to award bonuses to their staff above the established maximum, for introducing new technology, broadening the product mix and for increasing the output of export goods or import substitutes.[46] In short, in order to make new technology bonuses less marginal, a greater degree of discretion was permitted. The main problem, however, continued to lie in the artificiality of these incentives in comparison with the real pressures of the market (the 'invisible foot', as Joseph Berliner has called it). From 1975–82 there was a 50 per cent increase in the general sum of bonuses paid out for the introduction of new technology into the economy; within this overall increase price increments, supposedly justified by increased economic return, grew especially rapidly. Yet despite this *apparent* increase in quality and value to the customer, reflected also in the award of the official quality mark (*znak kachestva*) to a larger number of products, the *actual* rate of growth of economic return during this period was at best half that of the rate of growth of bonuses.[47] The payment of material rewards for innovation had become a kind of institutionalized confidence trick. Academician Trapeznikov noted caustically that the claim that by the end of the eleventh Five-Year Plan certain ministries would have 45 per cent of their output in the so-called 'highest category', equal or superior to the best foreign models in terms of quality and performance, was simply not credible.[48] The more generous bonuses for new technology were doubtless received with pleasure but appear to have made little difference to the traditional attitudes of ministries and enterprises towards their pattern of priorities in plan fulfilment. *Val* continued to reign supreme. The rosy vision of branch research institutes acting as catalysts, constantly challenging the conventional wisdom of producers and exposing the complacency of their superiors, conflicts sharply with the real tendency for 'leading ministeral officials [to] regard them only as administrative entities whose initiatives [could] be waved aside'.[49] Indeed, G. Marchuk, the Chairman of SCST has claimed that the enthusiasm for substantial innovation as distinct from minor modifications, may even have receded further:

In recent years there has been a decline in the responsibility of ministries, departments, associations and enterprises for the quality of work involved in drawing up and implementing plans for the creation, assimilation and utilisation of new technology. These plans increasingly include fewer significant and economically effective measures and the tasks set out in them are not placed under the same strict control as targets for volume of production, product assortment and other basic indicators.[50]

One is constantly struck by the central paradox which the Soviet R and D system presents: on the one hand the unswerving political commitment to technical progress, taking material form in the provision of massive resources, and on the other hand, the ambivalent and contradictory commitments of ministries and their subordinates, which collectively undermine the realization of these grandiose political objectives.

The innovation problem certainly does not lie in the overall scale of the Soviet R and D effort, though in some ways the pattern of resource allocation within this overall effort does raise a number of problems. The rate of growth of Soviet R and D personnel began to exceed that of the USA during the 1960s, and by the end of that decade the crossover point had occurred in the total numbers employed; this differential growth continued into the first half of the 1970s but was not sustained thereafter. By the late 1970s the number of scientists and engineers employed in R and D in the USSR (excluding the social sciences and humanities) was 828,160 compared with 595,000 in the USA.[51] In the strictly industrial sphere there was rough parity: 457,000 industrial R and D personnel in the Soviet Union compared with 390,000 in the USA.[52] It has become clear, however, that this rapid expansion of the late 1950s and 1960s was in some ways an optical illusion. New institutions were created in a formal sense across a very broad front, but in several cases they were Potemkin villages, deficient in basic equipment and supplies. This 'scattering of resources', which parallels a similar tendency in the industrial investment sphere, has had a long-term influence on styles of work. Even today, there are prestigious research institutes which are divided between many different premises in different areas of the major cities and lack even sufficient desks and bench space for their researchers.[53] Many research staff are obliged to work at home and put in only a fleeting appearance at their institutes. Complaints continue to abound about the lack of testing facilities and the absence of sophisticated equipment for computer-aided design.[54] It is little wonder, therefore, that in their isolated working environments researchers and designers often lack the practical skills that come from industrial contact or hands-on familiarity with large-scale equipment. A further factor undermining morale is that during the last 20 years the salaries of industrial R and D personnel have declined relative to other groups in industry.[55] This has given rise to rapid job turnover and a reduction in the flow of talented new entrants to technological institutes (*Vuzy*). Together, these factors have weakened the central links in the research–production chain: product development and plant design.

Another factor, which has a crucial bearing on the rate of technical progress in the Soviet economy and has been underplayed in previous Western studies is the scale and pace of industrial investment. This governs the rate of *diffusion* of new technologies, which are 'embodied' in successive vintages of capital equipment. During the 1950s capital investment grew on the average by 12.6 per cent per annum. By the 1970s this figure had come down to 3.6 per cent.[56] The original 1981–5 plan envisaged a 12–15 per cent increase over the whole planning period but this was soon revised downward by Brezhnev in November 1981 as a result of competing claims on resources, stemming principally from the implementation of the 'Food Programme'. Capital investment is currently growing at about 2 per cent per annum,[57] less than the planned rate of growth of national income. The above trends must have contributed to a slackening in the rate at which the results of research and development are introduced and diffused throughout the economy and may partly explain why the rate of economic return on each ruble spent on science and technology has actually fallen during the last ten years.[58] However, this argument needs to be taken further because there is strong evidence to suggest that official figures on capital investment are systematically distorted by concealed inflation. As a result of the lax granting of special price increments (*nadbavki*) for new items of machinery and equipment – a phenomenon referred to earlier – value estimates of new additions to capital stock grossly exaggerate the *real* growth in productive power.[59] This is an important mechanism of inflation in the Soviet centrally-planned economy. Fal'tsman has estimated that general price inflation of domestic (and foreign) equipment grew by 32 per cent over the 1976–80 period, exceeding the rate of growth of capital investment and thus signifying an absolute decline in the commissioning of productive capacity.[60] Given our knowledge of how the Soviet R and D system operates, one does not have to go as far as Fal'tsman to see what impact this sort of slowdown in investment would have on the behaviour of its constituent institutions. One can begin to perceive the main phases of a vicious circle, which would be extremely difficult to break out of:

(1) A slower growth in real investment leads to a slowdown in the rate of diffusion of new technology.
(2) This, in its turn, reduces the opportunities for practical familiarization with new technologies ('learning by doing') and at the same time intensifies pressures on the lower levels of the system for fulfilment of the production plan, which is geared to unrealistic formal expectations of productive capacity at higher levels.
(3) The squeeze on resources hits especially the 'soft' areas of full-scale experimental and testing facilities, which are essential for translating new ideas into future industrial practice.
(4) Political pressures from the centre for more rapid technical progress continue unabated without a significant improvement in the material environment which would encourage it.
(5) Scientists and designers respond formalistically with claims about the economic benefits to the economy of poorly proven equipment, which turn out later to be grossly exaggerated.
(6) This process undermines the quality of future capital investment and so the spiral continues.

CONCLUSION:
THE OPTIONS FACING SOVIET POLICY-MAKERS

The foregoing assessment of the weaknesses of industrial innovation in the USSR would not be altogether alien to many Soviet economists on the strictly empirical plane, though important differences in general interpretation would remain. This divergence of approach comes out in a particularly interesting way in Soviet writings about Western 'bourgeois falsifiers' – a category of literature which raises intriguing questions as to whether its main purpose is to criticize Western views or give them wider circulation! It emerges from these writings that there is a large measure of agreement between Soviet and Western writers on details relating to technological backwardness in some industries and about the organizational problems that hold back the rate of technical progress. There is also a mutual appreciation of the historic nature of the institutional transition which needs to be made in order to overcome these problems. When it comes to the *principle* of central planning, however, and its fundamental superiority over the market, a sharp line is drawn. The 'Scientific and Technical Revolution' (STR) poses new and unforeseen problems for all social systems and calls into question the adequacy of their institutional frameworks. However, the official Soviet line is that the contradiction between base and superstructure in the USSR is essentially non-antagonistic in character and can be overcome by administrative reconstruction, whereas the endemic crises and increasing technological unemployment, which are a growing feature of advanced capitalism are thought to be symptomatic of much deeper and more fundamental ills.[61] While it may be true that at the present stage of development an East–West

Table 1.3 Major reform options for speeding up technical progress in the USSR

Reform variant	Specific features	Flaws or problems of implementation
(A) Hungarian-type economic reform	A muted form of market socialism. Greater spontaneity and dynamism as a result of some administrative decentralization, greater competition and a liberalisation of the price system and foreign trade.	(1) Emergence of some of the characteristic economic problems of capitalism within the framework of state socialism. (2) Opposition of defence industry lobby whose resource priority depends on maintenance of central planning, and of middle-level party bureaucrats whose power depends upon administrative controls over resources. (3) Danger of inflamed class antagonisms as a result of widening income differentials.
(B) Technocratic reform	Streamlined central planning, which involves the further development of: science-production associations, more stringent state standards, ambitious inter-branch programmes of technical development, etc.	(1) Problems of transplanting administrative arrangements which have been successful in the defence industries into a civilian environment without guaranteed priority over resources or close customer participation. (2) This type of reform *administers* change but provides no incentive to *initiate* it in the first place. It conceptualizes the problem in terms of process rather than content.
(C) Enhanced discipline (neo-Stalinism)	Greater penalties for lateness, absenteeism and drunkenness; socialist competition between enterprises; more insistence on personal responsibility for failures and dismissal and demotion of incompetent managers; elimination of corruption.	Could bring some short-run improvement in *effort* but contains in a modified form the major flaws of the Stalinist system, which successive leaders have been trying to eradicate since 1953: (1) No stimulus to spontaneous initiative and creativity.

Table 1.3 continued

Reform variant	Specific features	Flaws or problems of implementation
		(2) No solution to the problem of the coordination of complex interdependent elements of an advanced technological society and, hence, (3) Misrepresentation of systemic problems as the fault or the failures of individuals.
(D) CMEA co-operation	Greater technological specialization and integration as a result of the rejuvenation of existing mechanisms of co-operation within the CMEA.	(1) Poor track-record for developing and disseminating advanced technologies. (2) Begs the question of institutional reforms in each country which would provide an initial *source* of innovation on which co-operation could be based. (3) Given the persistence of departmental barriers *within* countries, the possibility of harmonious co-operation *between* them seems implausible.
(E) West–East technology transfer	The selective acquisition of equipment and know-how from the most advanced Western countries in key areas of technological backwardness (e.g. electronics and computers); improvement of economic links with the West through the medium of co-operation agreements.	(1) Economic impact of imported technologies restricted by institutional obstacles to downstream diffusion. (2) Hard currency constraints in the light of the East European debt crisis and falling Soviet oil output. (3) The tightening of Western export controls and Soviet fears of political leverage.

technology gap exists in many industries, Western writers are accused of 'absolutizing' this gap and failing to see it in its long-term dialectical perspective.[62] Western writers are also taken to task for their simplistic assumption that innovation is a random occurrence which springs spontaneously from the market. This, again, is regarded as an 'absolutized' notion[63] which fails to take sufficient account of the fact that the growing scale and expense of science has to a great extent transformed it into a planned and directed activity in all countries; moreover, technological innovations are seen to rest ultimately on earlier advances in 'big science'. Of course, these Soviet writings raise enormous historical and empirical issues which can not be resolved here. But they also point in a very interesting way to the philosophical boundaries which could influence the Soviet approach to restructuring its R and D system in order to accelerate the pace of technical progress. In contrast to a widespread Western view that innovation rests upon the *raw dynamism* of the market, which must be controlled and directed in order to achieve a socially desirable outcome, the starting point for many Soviet writers is the elaborate apparatus of control and direction, created in order to impose a national interest from the centre, into which one hopes to inject dynamism. Thus, substantial agreement over many empirical details can still mask a fundamental clash in theoretical conception.

At the time of writing, the USSR, with its new General Secretary in command, is facing a number of critical choices. The experience of the last decade has demonstrated that minor tinkering with the economic system has not been enough to induce the profound psychological change, which would pave the way for more rapid technical progress. This much seems clear and is appreciated by a growing number of Soviet managers and officials.[64] But that still leaves a variety of options which might be pursued by the new leadership as it approaches its next Party Congress. These options, which are not mutually exclusive, are summarized briefly in table 1.3, together with the typical problems associated with each of them. They are discussed in greater detail in later chapters of this book by David Dyker (options A–C), Vladimir Sobell (option D) and Gary Bertsch (option E).

In December 1984, shortly before he took office, Mikhail Gorbachev addressed a special ideological conference in the Kremlin.[65] The main theme of his speech lay in its emphasis on the need for profound economic (and political) reform, a task which he equated in difficulty and importance with the heroic industrialization drive of the 1930s. This theme was echoed in a keynote article in *Kommunist* by the ailing Chernenko, which in its thrust and tone bore all the hallmarks of Gorbachev: especially in its espousal of the reformist notion that a contradiction existed between the forces of production and specific institutional forms. The next Party Congress, Chernenko (or Gorbachev?) predicted, would not be merely a chronological watershed between one five-year period and another, because 'our economy has now approached in earnest the frontier at which *qualitative* changes in it have become ... an imperative necessity'.[66] Thus, institutional reform is very much in the air. The specifics have yet to be announced. In the near future, however, we are likely to obtain some fascinating insights into the process of political decision-making in the USSR and into the broad outlook of the new leadership as it weighs on the one hand the economic pressures for reform, and on the other, the range of political consequences entailed in pursuing it.

APPENDIX TO CHAPTER 1

Table 1A.1 Invention: number of prototypes of new machines, equipment, apparatus and instruments

	1961–5	1966–70	1971–5	1976–80	1981	1982	1983	Average annual growth rates (on basis of previous aggregate period)
Total	23,178	21,272	20,006 (16,595)[a]	18,521 (17,523)	3,244 (3,602)	3,451 (3,608)	3,630 (3,866)	1971–5 = −1.19% 1976–80 = −1.48% 1981–3 = −1.42%
Machines and equipment (mainly in traditional industries – electric power, metallurgy, chemicals, machine tools, etc.)	16,626	15,560	15,190 (9,130)	13,971 (9,332)	2,465 (2,023)	2,623 (1,389)	2,830 (1,662)	1971–5 = −0.48% 1976–80 = −1.60% 1981–3 = −1.11%
Instruments, means of automation and computer technology (i.e. more advanced science-based industries)	6,552	5,712	4,816 (1,894)	4,550 (2,744)	779 (483)	828 (809)	800 (1,081)	1971–5 = −3.14% 1976–80 = −1.15% 1981–3 = −2.37%

[a] The figures in brackets refer to new products actually *introduced* into industrial production for the first time. Curiously, the pattern captured by the table diverges quite markedly. Whereas in the traditional sectors, the growth of both prototypes and new products is consistent, the rate of growth of new advanced products far exceeds that of prototypes in the equivalent industrial sectors. There appears to be no obvious explanation for this phenomenon.

Source: Data for all tables in this appendix taken from *Narodnoe Khozyaistvo SSSN*, various years.

Table 1A.2 Innovation: expenditure on the introduction of new technologies into industry and their economic impact

	1970	1971	1972	1973	1974	1975	1976	1977	1978	1979	1980	1981	1982	1983	Average annual growth rates (based on year immediately preceding period in question)
Number of measures introduced (1,000s)	423	449	497	509	579	621	633	670	675	725	773	753	780	795	1971–5 = +9.36% 1976–80 = +4.89% 1981–3 = +2.32%
Annual expenditures (milliard rubles)	5.01	4.90	5.33	6.56	6.71	7.52	7.98	8.17	8.66	8.88	9.70	9.90	11.70	11.3	1971–5 = +10.02% 1976–80 = +5.80% 1981–3 = +5.27%
Number of workers provisionally released (1 000s)	399	447	482	553	588	576	542	543	558	592	555	510	450	479	1971–5 = +8.87% 1976–80 = −0.71% 1981–3 = −2.41%
Actual economic effect from new technology (milliard rubles)	2.61	2.75	3.00	3.54	3.70	3.83	3.98	4.20	4.34	4.48	4.8	4.4	4.8	5.1	1971–5 = +9.35% 1976–80 = +5.06% 1981–3 = +1.86%

Table 1A.3 Diffusion: modernization of industrial processes at enterprises in different branches of industry (1,000 units)

	1966–70	1971–5	1976–80	1981	1982	1983	Average annual rate of growth (on basis of previous aggregate period)
Total	675	732	812	157	150	146	1971–5 = +1.69% 1976–80 = +2.19% 1981–3 = −1.40%
Ferrous metals	12.0	11.8	15.0	3.6	2.6	3.7	1971–5 = −0.33% 1976–80 = +5.42% 1981–3 = +2.00%
Chemicals and petrochemicals	18.8	34.5	35.3	6.9	6.2	6.2	1971–5 = +16.70% 1976–80 = + 0.46% 1981–3 = − 1.77%
Machine-building and metal-working	165	166	177	32	32	33	1971–5 = +0.12% 1976–80 = +1.32% 1981–3 = −1.73%
Construction materials	9.7	18.0	22.1	3.8	3.7	3.8	1971–5 = +17.11% 1976–80 = +4.55% 1981–3 = −2.96%

Table 1A.4 Incremental improvements: inventions and rationalization measures introduced into the national economy

	1970	1971–75 aggregate	1976	1977	1978	1979	1980	1981	1982	1983	Average annual growth rates (based on year immediately preceding period in question)
Number of inventions and rationalization suggestions utilized in production (1,000s)	3,414	18,583	4,030	3,988	4,014	4,019	4,048	3,974	3,986	3,969	1971–5 = +3.30% 1976–80 = +0.36% 1981–3 = −0.65%
Economic effect from utilization of inventions and rationalization measures (milliard rubles)	3.01	19.6	4.91	5.29	5.88	6.26	6.9	6.9	7.0	7.0	1970–5 = +11.96% 1976–80 = +8.69% 1981–3 = +0.48%
Expenditures on inventions and rationalization suggestions (milliard rubles)	229	1,404	331	332	352	359	379	392	395	n.a.	1970–5 = +7.86% 1976–80 = +3.76% 1981–2 = +2.11%

Table 1A.5 Rates of growth of labour productivity by branches of industry (1940=100)

	1960	1965	1970	1975	1976	1977	1978	1979	1980	1981	1982	1983	Average annual growth rates (based on year immediately preceding period in question)
Industry as a whole	296	372	492	657	679	706	732	749	769	789	806	835	1971–5 = +6.71% 1976–80 = +3.41% 1981–3 = +2.86%
Ferrous metals	265	335	415	529	550	556	572	573	571	569	569	576	1971–5 = +5.49% 1976–80 = +1.59% 1981–3 = +0.29%
Chemicals and petrochemicals	453	630	908	1,344	1,422	1,501	1,565	1,586	1,649	1,726	1,770	1,862	1971–5 = +9.60% 1976–80 = +4.54% 1981–3 = +4.30%
Machine-building and metal-working	472	645	938	1,419	1,517	1,619	1,724	1,824	1,917	2,009	2,089	2,200	1971–5 = +10.26% 1976–80 = +7.02% 1981–3 = +4.92%
Construction materials	365	516	678	896	917	935	950	939	946	963	970	1,006	1971–5 = +6.43% 1976–80 = +1.12% 1981–3 = +2.11%

2
The Civilian Production of the Soviet Defence Industry*

JULIAN COOPER

Speaking at the Twenty-fourth Congress of the Communist Party of the Soviet Union (CPSU) in March 1971, Party General Secretary L. I. Brezhnev revealed that 42 per cent of the volume of production of the defence industry served civilian purposes.[1] This statement received considerable attention in the West and may also have surprised many Soviet citizens. It was followed by a flurry of articles in the Soviet press by ministers of branches of the defence industry providing details of some of their civilian activities.[2] In retrospect it is curious that this public discussion of a topic normally shrouded in secrecy did not prompt any Western research into the nature and scale of the civilian production of the Soviet defence industry.[3] In the absence of such research, Brezhnev's reference to '42 per cent' has continued to puzzle Western scholars and, it also appears, the intelligence agencies.[4]

Apart from the vexed question of the real meaning of Brezhnev's remark, study of the civilian side of the defence industry would appear to have wider significance. If many enterprises of the defence sector manufacture civilian products on a regular basis, study of this activity may illuminate the functioning of the defence industry and its relationships with the civilian economy. It may also cast light on the question of the comparative technological level, quality and efficiency of the defence and civilian sectors, and the possibility that the defence industry may be able to assist in raising the technical standards and productivity of the economy as a whole. Finally, the investigation may contribute to an understanding of the possibilities and problems of conversion from military to civilian production in the event of such conversion becoming a practical possibility. It was interest in this latter issue that initially prompted the present research.

THE SOVIET DEFENCE INDUSTRY

For the purpose of this study, the defence industry is considered to be that part of Soviet industry administered by the nine ministries of the defence industry group. These ministries comprise:

*This study was undertaken as part of a research project on 'The Economic Potential of the Soviet Engineering Industry', funded by the Economic and Social Research Council (ESRC). The support of the Council is gratefully acknowledged.

Ministry of Medium Machine-building (Minsredmash, MSM).
Minister: E. P. Slavskii, appointed 1957.
Principal products: nuclear weapons, uranium mining and processing.

Ministry of General Machine-building (Minobshchemash, MOM).
Minister: O. D. Baklanov, appointed 1983.
Principal products: missiles, space rockets and vehicles.

Ministry of the Aviation Industry (Minaviaprom, MAP).
Minister: I. S. Silaev, appointed 1981.
Principal products: military and civilian fixed-wing aircraft and helicopters.

Ministry of the Defence Industry (Minoboronprom, MOP).
Minister: P. V. Finogenov, appointed 1979.
Principal products: ground forces equipment – tanks, armoured vehicles, artillery, small arms and optical equipment.

Ministry of Machine-building (Minmash, MM).
Minister: V. V. Bakhirev, appointed 1968.
Principal products: conventional ammunition, explosives and fuse mechanisms.

Ministry of the Ship-building Industry (Minsudprom, MSP).
Minister: I. S. Belousov, appointed 1984.
Principal products: military and civilian ships and boats.

Ministry of the Electronics Industry (Minelektronprom, MEP).
Minister: A. I. Shokin, appointed 1961.
Principal products: electronic components and a rapidly widening range of electronic end products.

Ministry of the Radio Industry (Minradprom, MRP).
Minister: P. S. Pleshakov, appointed 1974.
Principal products: radio and radar equipment, computers.

Ministry of the Communications Equipment Industry (Minpromsvyaz', MPSS).
Minister: E. K. Pervyshin, appointed 1974.
Principal products: communications equipment, consumer radio and television equipment.

These ministries share certain common features distinguishing them from the rest of Soviet industry. Unlike the civilian ministries, their annual growth of output is not separately reported in the published statistical reports of the Central Statistical Administration. There is a distinct Party Central Committee Department of the Defence Industry (headed by O. S. Belyakov); the activities of the 11 civilian machine-building ministries are supervised by a separate Machine-building Department, and of the chemical, petrochemical and microbiological industries by a Chemical Industry Department. Gosplan also has a distinct defence industry department under one of its first deputy chairmen; this appears to be specifically concerned with the activities of the nine above-named ministries, plus the military production undertaken by civilian ministries. General oversight and co-ordination of the activities of the defence industry ministries is realized through the Military–Industrial Commission of the Presidium of the USSR Council of Ministers, headed by L. V. Smirnov, a Deputy Chairman of the Council. In this chapter we are not concerned with the military production undertaken by enterprises of civilian industrial ministries: this is a topic virtually impossible to research on the basis of published Soviet sources.

HISTORICAL BACKGROUND

The Tula samovar symbolizes the historic link between military and civilian production in Russia: today the largest, best-known maker of samovars is still a munitions plant of Tula (the Vannikov 'Shtamp' factory) under the Ministry of Machine-building.[5] Before the Revolution many large arms plants manufactured civilian products alongside weapons – the giant Putilov works of St Petersburg (the present-day Kirov factory) built locomotives, wagons and excavators in addition to artillery systems. After the Revolution and Civil War, the civilian activities of defence industry factories revived: it was the Putilov works which pioneered volume tractor building from the mid-1920s, making a copy of the American 'Fordson'. Towards the end of the decade, policy for the future expansion of the defence industry was established. The head of the mobilization sector of the Red Army staff, S. Ventsov, called for a policy of 'military assimilation' (*assimilyatsiya*) having two dimensions: first, the use of spare capacity at military factories for the production of civilian items technologically related to the basic military products so as to maintain appropriate skills, and, second, the creation of conditions permitting the manufacture of military materiel at many civilian enterprises in the event of war. It was believed that this approach would permit an adequate expansion of arms production on the basis of a relatively small 'core' defence industry, while at the same time the defence sector would contribute to the industrialization of the country.[6] Such a policy was pursued during the 1930s, although the core defence industry appears to have expanded to a much greater extent than originally envisaged. By the late 1930s the range of civilian products made at defence industry enterprises was extensive, and for some items they were the sole producers. The goods made included machine tools (the superior skills of the defence industry allowed it to pioneer the domestic production of many technologically complex new models), textile machinery, tractors, turbines, railway wagons, excavators, optical equipment and consumer goods, including watches, samovars and gramophone records.

During the war, civilian production at defence industry enterprises virtually ceased, to be progressively resumed as the end of hostilities approached. In the immediate post-war period, many arms factories fully converted to civilian production, while those remaining in the defence sector contributed to the reconstruction effort. Some of the large artillery works made equipment for the oil industry; the giant Nizhnii-Tagil Uralvagonzavod, the largest tank producer of the war, resumed the volume production of railway freight wagons.[7] However, with the onset of the Cold War and the intensification of the Soviet drive to attain a nuclear capability, the defence industry's 'civilianization' appears to have suffered a reverse. In the late 1940s some large new enterprises of the civilian engineering industry were transferred to the defence sector, for example, the Dnepropetrovsk truck factory converted to the production of missiles.[8] From about 1953 there appears to have been a renewed emphasis on the need for defence industry enterprises to contribute to the civilian economy: it was in that year that the Dnepropetrovsk works began building tractors.[9] This turn was associated with the policy of accelerating the development of consumer goods industries initiated by G. M. Malenkov, then Chairman of the USSR Council of Ministers.[10] A substantial Party and government decree of October 1953 provided for considerable involvement of the defence industry in an ambitious programme for rapidly expanding the production of a wide range of consumer goods.[11] It was around this time that many enterprises of the defence sector initiated or consolidated their consumer goods production and in many cases they have retained it to this day. Malenkov was removed from his post in January 1955, and the priority of heavy industry reasserted, but the Soviet consumer can be grateful for his legacy. The policy of involving the defence industry in the satisfaction of consumer demands has been reaffirmed in successive decrees, usually under the standard formula that 'all branches of industry' should participate.

Since 1953 official policy has consistently favoured the involvement of the defence industry in civilian activities, and this view has received endorsement in Party and government decrees on the development of specific branches of the economy, which have included explicit references to the role of various defence industry ministries, for example in the manufacture of medical equipment, the improvement of consumer services, and the development of land reclamation.[12] In 1970, a few months before Brezhnev's statement at the Twenty-fourth Party Congress, measures were adopted to increase the contribution of the defence sector to the supply of agricultural equipment.[13] In 1981, at the Twenty-sixth Party Congress, Brezhnev called for greater involvement of the research facilities of the defence sector in efforts to improve the technological level of the light, food and medical industries, and, once again, agriculture.[14]

THE PRODUCTS

The range of civilian products manufactured by the defence industry is extremely diverse. Here space permits only a summary review, indicating the principal products, the ministries involved and the location of the main identified enterprises. These data are in the main unfamiliar to Western specialists, so it is worth quoting specific examples and sources of information.

Transport equipment

Besides civilian aircraft (MAP) and ships (MSP), enterprises of the defence industry produce a range of equipment for rail and road transport. Up to one-third of all rail freight wagons are built at the Nizhnii-Tagil works of MOP, the largest Soviet tank building enterprise.[15] Of the three enterprises which build tramcars, the largest volume producer is the Ust'-Katav works in the Urals under MOM.[16] In 1966, in order to expand the production of cars while the new 'VAZ' plant was under construction, the 'Izhmash' association of MOP organized the building of the 'Moskvich' car in parallel with the Moscow plant of the motor industry. This multi-product works at Izhevsk (now Ustinov) is one of the largest production associations of the Soviet engineering industry; its chief designer is M. T. Kalashnikov, creator of the famous assault rifle which bears his name.[17] All engines for the 'Moskvich' (both Moscow and Izhevsk variants) are built by the Ufa aeroengine works of MAP.[18] Minoboronprom enterprises also build motorcycles (Kovrov and Izhevsk) and scooters (Tula), while the two largest producers of bicycles are enterprises of Minmash (Perm' and Penza).[19] The aviation industry makes engines for motorized bicycles (Leningrad) and also snowmobiles (Rybinsk).[20] Some trucks are built outside the motor industry, possibly at enterprises of MOP.[21]

Agricultural equipment

The most significant contribution of the defence industry to the agricultural sector takes the form of the production of tractors. More than 300,000 'Kirovets' heavy-wheeled tractors have been built by the Leningrad 'Kirov factory' association since 1964,[22] while an enterprise of Minobshchemash (Dnepropetrovsk) has been building small, universal tractors ('YuMZ') since 1953.[23] Together they account for approximately 15 per cent of total Soviet tractor output.[24] Engines for the 'YuMZ' are supplied by the Rybinsk aeroengine works.[25] Several associations of the aviation industry are now organizing the manufacture of mini-tractors and cultivators for use on private plots and gardens (Ufa, Moscow, Zaporozhe).[26] Some trailers for use with tractors, and components for their production, are made by the ship-building industry (Nikolaev).[27] Seed drills and other agricultural machines are built by enterprises of Minmash; mineral fertilizer spreaders are made by MOP.[28] Enterprises of the aviation industry make hen batteries and other equipment for poultry farming, while the Rybinsk aeroengine works is a substantial producer of milk separators.[29] Since the early 1970s, the ship-building industry has been building large irrigation installations ('Fregat' and 'Kuban' – Leningrad, Pervomaisk), with aluminium tubing supplied by the Kuibyshev metallurgical works of the aviation industry.[30] Some irrigators are also supplied by enterprises of MOP.[31]

Industrial equipment

Since the early 1930s, enterprises of the defence industry have built machine tools on a regular basis for their own needs and for use in other branches of industry. Machine-tool building is today undertaken by enterprises of several defence industry ministries, including MOP (Izhevsk, Votkinsk), Minmash (Kuibyshev, Leningrad, Penza, Vladimir), MAP (Vladimir), MSP and MOM.[32] The Savelov 'Progress' association (Kalinin *oblast'*) is reported to produce up to one-fifth of all Soviet numerically-controlled machine tools; its ministerial affiliation has not been established, but it is not under the civilian machine-tool ministry.[33] Laser technological equipment is produced by the electronics industry and MEP also makes some control units for NC machine tools.[34]

Almost all the defence industry ministries now build industrial robots.[35] The ship-building industry makes NC plasma metal-cutting equipment, and industrial painting and welding equipment (Leningrad).[36] Lasers are made by enterprises and institutes of MOP and MEP; the former also supplies much industrial optical equipment.[37] Other products include chemical industry equipment (MOP), textile machinery (MM), food industry machinery (including a process for making dried milk supplied by Minsredmash), and diesel engines (MSP, MAP, MOP).[38] Petrol-engined saws for use in forestry are produced exclusively by an enterprise of Minmash (Perm'), with chains supplied by MOP (Izhevsk).[39] Minmash also builds excavators (Zlatoust), while dredges for the gold industry are a product of the massive Perm' (Motovilikh) Lenin works of Minoboronprom, a major centre of the Soviet artillery industry.[40]

Fuel and power industry equipment

For many years enterprises of the defence industry have supplied equipment for the fuel and power industry. Machinery for hydro-electric power stations is produced by MOP (Leningrad, Perm') and MSP (Leningrad).[41] Mobile power stations used by the oil and gas industries are made by the aviation industry (Zaporozhe) and floating power stations by the ship-building industry (Tyumen').[42] Enterprises of MOP produce equipment for the oil industry, including turbo-drills supplied by the Perm' Lenin works.[43] The ship-building industry is now building off-shore oil rigs (Astrakhan', Leningrad).[44] A number of defence industry enterprises have contributed to the gas pipeline building programme, supplying power units for gas compressor stations (MAP – Kuibyshev, Kazan, Perm'; and MSP), and radio-communications equipment (MPSS).[45] Minsredmash is actively involved in the civilian nuclear power programme, with responsibility for uranium mining and processing, and it probably supplies some equipment for nuclear power stations. Pumps and other equipment for the latter are also produced by the Leningrad Kirov factory of MOP.[46]

Chemical products

It is not easy to determine the contribution of the defence industry to the chemical industry, but the available evidence indicates that enterprises of Minmash produce some fertilizers and agricultural chemicals, industrial explosives, plastics, celluloid, and paints and dyes. It also makes some domestic chemical products.[47]

Metallurgical products

Some of the large arms factories had their own steel-making capacity before the Revolution and have retained it to this day. While much of the steel produced by MOP (including the Leningrad 'Kirov' and 'Bol'shevik', and Perm' Lenin enterprises) and MSP (Gor'kii 'Krasnoe Sormovo') is probably consumed within the defence sector, some may be used by the civilian engineering industry.[48] The aircraft industry has a number of plants producing non-ferrous castings and forgings, including the large Kuibyshev Lenin metallurgical works, which supplies many civilian customers with aluminium products, and has pioneered a number of new metal-forming processes.[49]

Computers

The only civilian ministry producing computers is the Ministry for Instrumentation and Means of Automation (Minpribor), which builds a wide range of control computers. The rest of the computer industry is subordinated to ministries of the defence industry group. The principal producer is Minradprom, building main-frame, universal models (Minsk, Penza, Kazan, Moscow, Erevan, Baku, Brest).[50] An enterprise of the radio industry is producing the first Soviet personal computer, the 'Agat'.[51] In recent years the electronics industry has emerged as a large-scale producer of computers, making mini and micro models (Leningrad, Kiev, Voronezh). Recently, Minpromsvyaz' has also entered this field (Riga, Kiev).[52] There is evidence that some computing equipment was produced by enterprises of Minoboronprom during the 1960s and early 1970s (Ulyanovsk?), although no recent reference has been traced.[53] At this time Minaviaprom also built some calculators and computing equipment (Kirov); again, no recent reference has been found.[54] Since the mid-1970s, hand-held electronic calculators have been mass produced at Minelektronprom enterprises (Voronezh, Leningrad, Kiev, Fryazino, Frunze, Ulyanovsk).[55]

Radio, electronic and communications equipment

The three defence industry ministries with principal responsibility for electronics-based products (MEP, MRP, MPSS) make a substantial contribution to the civilian economy. Minelektronprom appears to supply a very high proportion of all electronic components consumed throughout the economy.[56] It is now the major producer of microprocessors, although some are also made by enterprises of Minpromsvyaz' (Riga, Kiev).[57] Civilian products of Minradprom include radar and flight-control systems used by Aeroflot and systems employed by the merchant fleet, radio-measuring devices, radio telescopes and other equipment for radio astronomy.[58] Minpromsvyaz' is the principal manufacturer of telephone and telegraph equipment used throughout the economy, and other communications equipment of all types, including systems for radio and television, cable and fibre-optics networks, and satellite television and communications.[59] For these three ministries, civilian production must represent quite a sizeable share of total output. Other defence industry ministries also produce electronics-based equipment for civilian uses, for instance, Minsudprom makes hydro-acoustic fish-detection systems for the fishing fleet (Taganrog, Vladivostok).[60]

Medical equipment

The Ministry of the Medical Industry has basic responsibility for medical equipment, but almost all the defence industry ministries have some involvement. Their contribution appears to be focused on the supply of technologically complex apparatus incorporating laser, electronic, optical and X-ray systems. References to the participation of MAP, MOM, MOP, MSP, MEP, MRP, MPSS and Minsredmash (X-ray equipment) have been traced.[61]

Consumer goods

Enterprises of the defence industry make a substantial contribution to the production of consumer goods, with particular emphasis on electrical and electronic durables. The only ministry for which no explicit reference has been traced is Minsredmash, but this probably relates to the exceptional secrecy surrounding its activities. Since 1968, when the system was introduced, defence industry ministries have been designated 'head' organizations for certain consumer items with responsibilities for the overall technological level, quality and volume of production of the products concerned. Thus, Minmash is head ministry for samovars, being itself a major producer, while Minoboronprom leads the production of cameras and other consumer optical equipment.[62] Minaviaprom has responsibility for prams and aluminium kitchen utensils, both of which it manufactures in large quantities, Minpromsvyaz' is head ministry for a wide range of consumer radio and electronic goods, including televisions and tape recorders, while Minsudprom has responsibility for motor boats and yachts.[63] For most products, however, the head ministries are outside the defence sector – which gives rise to difficulties and occasional conflicts.

Of the principal consumer durables, refrigerators are manufactured on a substantial scale by Minobshchemash (Krasnoyarsk) and MAP (Saratov), but also by Minmash, Minoboronprom and Minradprom (Zlatoust, Yuryuzan', Murom, Orsk, Dnepropetrovsk).[64] Washing machines are produced by enterprises of MAP (Arsen'ev, Ulan-Ude, Kiev, Voronezh), MSP (Gor'kii, Komsomol'sk-na-Amure), MM and MOP (Cheboksary, Kopeisk, Votkinsk, Vysogorsk).[65] The aviation industry is a substantial producer of vacuum cleaners (Dnepropetrovsk, Moscow).[66] Gas and electric cookers are built by enterprises of Minmash (Zlatoust) and MOM (Dnepropetrovsk, Voronezh?).[67] Almost all televisions and radios are produced by the defence industry ministries. The majority of the former are manufactured by more than 20 enterprises of Minpromsvyaz' and Minradprom, but some are also made by Minobshchemash (Kiev, Khar'kov),[68] and small, portable sets are now made by enterprises of the electronics industry, the supplier of TV tubes (Aleksandrovsk, Khar'kov, Khmel'nitskii, Leningrad, Moscow, Vitebsk).[69] The production of radios, record players and tape recorders is more widely diffused: at least seven ministries are involved, with some of the best models being produced by the aviation and ship-building industries. Several ministries are now engaged in the manufacture of video equipment (MEP, MRP, MAP, MM), although the scale of production remains limited.[70] Minoboronprom is the principal producer of consumer optical equipment: cameras (Leningrad, Krasnogorsk, Minsk, Kiev), binoculars (Zagorsk), telescopes (Novosibirsk), although some cameras are made at a Khar'kov factory of the aviation industry, and projection equipment and some cine cameras by enterprises of the ship-building industry (Kiev).[71] Most Soviet clocks and watches are produced by Minpribor (the head ministry), but defence industry producers include Minmash (Kuibyshev, Vladimir) and Minaviaprom, which makes alarm clocks (Rostov).[72] Digital electronic watches are a monopoly of the Minsk 'Integral' association of the electronics industry.[73] Many enterprises of the defence industry produce furniture, glass consumer goods, electric irons, hairdryers, light fittings, kitchen utensils, umbrellas and toys. Electric razors are manufactured by several ministries, including MOM, MAP and MM.[74] At least three ministries are involved in the building of motor boats and other pleasure-craft (MSP, MAP, MOM); motors for them are supplied by several enterprises of the aviation industry (Kazan, Kuibyshev, Moscow, Perm').[75] As this summary review indicates, there must be very few consumer goods *not* produced within the defence sector.

THE SCALE OF CIVILIAN PRODUCTION

It is difficult to obtain even an approximate estimate of the overall scale of the civilian production of the nine defence industry ministries and its change over time. Some fragmentary evidence is available on the civilian activities of individual ministries; and it is possible to assemble statistics on the volume of production of specific products and the proportion of total Soviet output accounted for by the defence sector. Brezhnev's '42 per cent' is relevant to the former. It is the author's belief that this proportion refers not to the whole defence industry, but to a single ministry, namely, Minoboronprom – the Ministry of the Defence Industry. This view gains support from additional evidence supplied by the economist, Subbotskii. In a work published in 1979, he observed that: 'In the ninth Five-Year Plan 42 per cent of the total volume of production of the given branch [i.e. *oboronnaya promyshlennost*' – JMC] went to civilian purposes. This includes oil drilling installations, machine tools, railway wagons, tractors, motor vehicles, optical equipment, refrigerators, kitchen utensils, etc.'[76] All these products are manufactured by enterprises of Minoboronprom, and the particular product range specified by Subbotskii applies to this ministry alone. Of the defence industry ministries, MOP does appear to have the widest range of civilian products, and it may well be the ministry with the highest share of civilian output; hence the choice of this most impressive example by Brezhnev. Other ministries probably have somewhat smaller shares of civilian output, although in some cases the proportions may still be quite substantial; for instance, in the aviation, ship-building and communications equipment ministries. For the aviation industry there is a CIA estimate that between 1967 and 1977 in ruble-cost terms, one-third of all aircraft built were for civilian purposes. In the ship-building industry, civilian use accounted for less than one-third of all ships and boats.[77] For only one ministry has a ruble figure for civilian production been found: according to the ninth Five-Year Plan (1971–5), Minaviaprom was to produce civilian goods (excluding aircraft) to a value of 3,000 million rubles; in 1981–85, 10,000 m.r., giving an annual average rate of growth of almost 13 per cent.[78] For some ministries there is information on the scale of consumer goods production. The largest manufacturer is Minpromsvyaz', with a 1980 planned output of more than 3,400 m.r.; in 1982

Minelektronprom's production was 1,400 m.r. and Minsudprom's 613 m.r.[79] In 1970 the aviation industry made consumer goods to a value of 307 m.r., rising to 945 m.r. according to the 1980 plan, giving an annual average rate of growth of almost 12 per cent.[80] Compared to the consumer goods output of the civilian engineering ministries, these are substantial totals. The specialized producer of consumer durables, the Ministry of Machine-building for the Light and Food Industries (Minlegpishchemash) had an output of cultural and household goods of 2,000 m.r. in 1983, the machine-tool industry just over 400 m.r. and the heavy engineering industry 200 m.r.[81]

Table 2.1 presents data on the defence industry's share of the total Soviet production of a range of products for the period 1965 to 1980. The list of products is far from complete and determined solely by the availability of appropriate statistical information. In some cases the indicator employed, share of production in physical unit terms, leads to an underestimation of the defence industry's contribution; for example, for tractors, a share indicator in terms of horse-power would give a much higher proportion. Nevertheless, the table provides an indication of the scope and scale of the contribution of the defence industry to the civilian economy, and suggests that its share has been quite stable over time, showing a modest tendency to decline over the 15-year period. This may be misleading, in so far as the table does not include a number of new products assimilated in recent years, for example: industrial robots, irrigators, pipeline equipment and medical apparatus. In the case of consumer durables, the reduced share can be explained at least in part by the efforts of the head ministry, Minlegpishchemash, to raise the level of specialization of production and close down small-scale manufacture at enterprises of other ministries.[82] While the defence industry share of the output of bicycles, motorcycles and mopeds has declined somewhat, in absolute terms it has continued to exhibit a steady upward trend. With additional research it should be possible to extend the range of products to include a range of industrial equipment and chemical products.

THE ORGANIZATION OF CIVILIAN PRODUCTION

Overall responsibility for civilian production in branches of the defence industry appears to rest with one of the deputy ministers of each of the ministries concerned. It is frequently these deputy ministers who write in the press and answer occasional criticisms of inadequate quality.[83] The aviation industry has a specialized Main Administration for Sales and Civilian Production, created in the late 1960s, and its head (currently A. A. Ryumin) is the usual author of press articles on the civilian activities of the ministry.[84] In other ministries it appears that the sales administration has responsibility for consumer goods production, and five of the ministries have created central commercial and advertising organisations (TsKRO) for handling relations with the trade system and sales promotion ('Elektronika', 'Radiotekhnika' (MRP),

Table 2.1 The share of total Soviet output of civilian products from enterprises of the defence industry (per cent of total output in physical unit terms, unless otherwise specified)

Product	1965	1970	1975	1980
Tractors (MOP, MOM)	13	13	13	15
Railway freight wagons (MOP)	n.	c. 33	c. 30	c. 27
Tramcars (MOM)	70	55	63	60
Passenger cars (MOP)	—	10	10	10
Motorcycles (MOP)	68	63	61	n.
Motor-scooters (MOP)	100	100	100	100
Bicycles (MM)	44	37	c. 33	c. 30
Mopeds and motorized bicycles (MM)	23	20	21	21
Metal-cutting machine tools (MAP, MOP, MM, MOM, MSP)[1]	n.	(10)	(10)	(10)
including NC machines (MAP, MOP, MOM)[1a]	n.	(40)	n.	(25)
Crude steel[2] (MOP, MSP)	12	10	9	n.
including alloyed steels[2] (MOP)	29	25	29	n.
including electric-arc steel[2]	54	53	55	n.
Refrigerators (MAP, MOM, MM, MOP, MRP)	53	54	51	47
Washing machines (MAP, MM, MSP, MOP)	40	37	c. 35	c. 35
Vacuum cleaners (MAP)	57	45	c. 33	c. 33
Tape-recorders (MEP, MRP, MPSS*, MAP, MSP, MM)	c. 90	c. 90	c. 90	c. 90
Television sets (MPSS*, MRP, MOM, MEP)	100	100	100	100
Radios (MPSS*, MRP, MEP, MSP)	100	100	100	100
Video-recorders (MEP, MAP, MSP, MM)	100	100	100	100
Clocks and watches (MM, MAP, MEP)	12	12	11	10
Cameras (MOP*, MAP)	100	100	100	100
Furniture (all?) (value terms)	n.	n.	n.	c. 2
Domestic chemical products (MM) (value terms)	n.	c. 3	c. 3	n.

n. not known; c. considered to be reasonably accurate estimate.
[1] Estimate based on known proportion built by the Ministry of the Machine-Tool Industry.
[1a] As 1, but more reliable as less produced by other civilian ministries.
[2] Proportion shown is that produced outside the Ministry of Ferrous Metallurgy; overstates share of defence industry as some steel is produced by civilian engineering ministries, notably heavy and power machine-building.
* head ministry for the consumer good.

Note: This table is provisional and subject to refinement in the light of additional research. It has been compiled from Soviet sources, too numerous to cite here, including press references to the output of individual enterprises, and republican, regional and city statistical handbooks.

'Orbita' (MPSS), 'Rassvet' (MOP) and the aviation industry).[85] These organizations are also responsible for the creation of networks of 'firm' shops under the same names, providing them with direct retail outlets and closer links between producing enterprises and customers; for example, cities with 'Elektronika' shops now include Moscow, Leningrad, Kuibyshev, Saratov, Sochi, Tbilisi, Tomsk, Voronezh and Pskov.[86]

At the level of the enterprise, the forms of organization of civilian production are diverse. In some cases, enterprises work fully, or predominantly, for non-military customers on a more or less permanent basis. This appears to apply to the aviation, ship-building and communications equipment ministries. The production association provides a convenient means of combining factories of large-scale, specialized civilian production with enterprises pursuing the basic, military-related activity of the association. Minoboronprom provides good examples: the tractor-building of the 'Kirov factory' association and the motor cycle, car and machine-tool building of 'Izhmash'. A related, but somewhat lower form of specialization is the organization within enterprises of specialized production shops and sectors for civilian and consumer goods. This solution has become more widespread in recent years as it offers the possibility of employing specialized, high-productivity production equipment permitting higher quality and lower production costs. The aviation and ship-building ministries have been particularly active in organizing such specialized facilities for consumer goods production: the former planned to create 12 new shops during the 1971–5 plan period, while the latter organized 11 during the following five years.[87] The Rostov helicopter works, for example, in 1971 created a shop employing 300 workers, engineers and designers – previously all consumer goods had been made in the basic shops of the enterprise.[88] An additional advantage of the specialized factory or shop may well be that the personnel concerned do not require security clearance or, at least, have less stringent security conditions than those employed in the basic, military, production units.[89]

A less specialized form of production, but one widely employed, is so-called 'assimilation' (*assimilyatsiya*) referred to at the beginning of this chapter: the use of the same production facilities for making both the basic product and civilian goods.[90] Assimilation can take a variety of forms. In some cases, the civilian product is manufactured from beginning to end in the same production shops as the basic product using the same equipment and personnel. In other cases, machining and assembly processes are separate, but draw on common preparatory facilities (foundry and forge-press shops). Many intermediate variants are possible. Assimilation provides a means of employing reserve capacity which could be switched to militarily-related production in the event of need. In the case of the aviation industry, Soviet writers have explicitly acknowledged that reserve capacity is used in this way.[91] Unfortunately, there is virtually no information on the relative proportions of the different forms of organization of civilian production. The only data refer to the total Soviet production of electrical domestic appliances (in which the defence sector plays a substantial role). It is estimated that 20 per cent of the total volume of production derives from specialized enterprises; 15 per cent from specialized production shops of enterprises making other basic products; 25 per cent from enterprises having specialized shops for final assembly; and 40 per cent is produced on an assimilation basis in basic production shops.[92] Nevertheless, the evidence strongly suggests that the trend of recent years has been towards a greater specialization of civilian production at defence industry enterprises. It is frequently stressed that this solution permits the use of specialized production equipment and this would seem to imply that the requirement of rapid convertibility has been relaxed. While it is probably the case that much of the capacity of the defence industry devoted to civilian purposes is regarded as potentially available for military production (some at relatively short notice, providing a surge capability), it may now be incorrect to assume that all such capacity is regarded as a reserve for military purposes.[93]

There appears to be no standard pattern for administrative arrangements for civilian production within defence industry associations and enterprises. In some large associations there are directors responsible for the civilian activity; the Kirov works, for example, has a director of tractor production, and at 'Izhmash' there is a director of motor vehicle production.[94] Representatives of the aviation industry have stressed that it is the policy of the ministry to vest responsibility for civilian activities at a high level: a deputy director or the chief engineer of an enterprise answers for the organization of such production and its quality.[95] From the fragmentary evidence available it is not apparent that involvement with civilian production limits career advancement possibilities within the defence industry. The chief engineer of the Leningrad 'Kirov factory' association between 1976 and 1983, Yu. A. Sobolev, was formerly director of tractor production and before 1972 was director of automobile production at 'Izhmash'.[96] The chief engineer of 'Izhmash' at the time when the 'Moskvich' car was being introduced into full-scale production, Yu. D. Maslyukov, later became a deputy minister of Minoboronprom and is now a first deputy chairman of Gosplan.[97] The former head of refrigerator production at the Krasnoyarsk Lenin machine-building factory of Minobshchemash, V. Mikhailov, is now the general director of a large science-production association in Leningrad (not identified, but assumed to be in the defence industry).[98] Indeed, it may now be the case that senior managerial personnel with experience of civilian production are considered especially well-qualified for advancement in the defence industry. They will probably have had more opportunity to travel abroad, may be better acquainted with the latest technological and managerial thought, and can be expected to be more cost-conscious than their colleagues who have only worked on military contracts.

Another feature of the civilian production of the defence sector is the existence of well-developed, inter-ministerial co-operative production practices. The tractor-building of the Dnepropetrovsk works has been described (by Brezhnev) as 'co-operative production',[99] and the building of the Izhevsk 'Moskvich' involves a large number of enterprises of different ministries of the defence industry and also the civilian sector. Car engines, clutches and gearboxes are supplied by the aviation industry (Ufa, Tyumen' and Omsk), steering, gear and light fittings by enterprises of Minmash (Perm' and Vladimir), springs and other components by Minoboronprom factories, including Votkinsk and

Perm'.[100] The available evidence indicates that defence industry enterprises making civilian products are by no means immune from supply problems, but complaints in the press are generally directed at ministries outside the sector.[101]

QUALITY AND TECHNOLOGICAL LEVEL

It is generally recognized that the quality of Soviet military materiel compares favourably with that of most products of the civilian economy. Enterprise quality control systems are directly reinforced by the presence of representatives of a powerful and demanding customer, the Ministry of Defence.[102] Given the general culture of production at defence industry enterprises, one would expect their civilian products also to be of relatively high quality, even though not subject to the control of the *voenpredy*. The available evidence tends to support this supposition, but also reveals an interesting pattern of ministerial and enterprise differentiation. Conclusions can be drawn from at least three types of evidence: (1) comment and complaints in the press; (2) information on the proportion of the output of ministries and enterprises judged to be of the highest category of quality and bearing the state mark of quality (*znak kachestva*); (3) information on the extent to which products are sold on foreign markets. On all three counts one ministry emerges as a consistent producer of high-quality civilian products of a relatively advanced technological level: Minaviaprom. Press comment is invariably favourable and its products are frequently singled out for praise, for example, the 'Saratov' refrigerator, which has the longest guaranteed service life of all Soviet models (three and a half years), the 'Raketa' vacuum cleaner (Dnepropetrovsk), and the Rybinsk tractor engine.[103] Some of the enterprises of the ministry have extremely high shares of output with the state quality mark: 100 per cent of the attested production of the Rybinsk aeroengine association (1983), 97 per cent of the attested goods of the Dnepropetrovsk aggregate factory ('Raketa' vacuum cleaners, 1980), 96 per cent at the Khar'kov 'FED' association ('FED' and 'Mikron' cameras, 1981), and 90 per cent of the refrigerators of the Saratov electro-aggregate works; a similarly high proportion of consumer goods manufactured by the Kazan helicopter factory (1983) were awarded the quality mark.[104] Representatives of the aviation industry claim that civilian products are usually manufactured with the same technological discipline and quality control procedures as used for the basic products.[105] This probably applies to Minobshchemash as well, but less evidence is available. The 'Biryuza' refrigerator is generally acknowledged to be the best Soviet model, with almost 100 per cent attested as the highest quality, the 'YuMZ' tractor is evidently a quality product (94 per cent of the attested production of the Dnepropetrovsk enterprise had the state quality mark in 1980), and the electric razors made by the ministry are regarded as the best produced in the country.[106] In 1979 more than half the consumer goods produced by MOM had the quality mark (40 per cent for Minpromsvyaz' and 28 per cent for MOP); but only 24 per cent of the output of the civilian electrical engineering industry and a mere 13 per cent of the specialized consumer durables produced by Minlegpishchemash achieved this level of official recognition.[107]

For other ministries of the defence industry the position is not always so favourable. Products of one ministry, Minmash, are frequently criticized for their inadequate quality and dated designs. Targets of criticism have included its refrigerators (in particular the 'Polyus' of the Zlatoust Lenin works), electric razors (Leningrad Kalinin works), 'Kometa' tape recorders (Novosibirsk) and the samovars of the Tula 'Shtamp' factory.[108] There are few complaints of poor quality for the products of Minoboronprom and Minsudprom (in 1983 half the latter's consumer goods production carried the quality mark),[109] but regular criticism of inadequate rates of product renewal.[110] Some enterprises of the former have very high shares of output attested to be of the highest quality, for example, the Tula arms factory with over 99 per cent and the Nizhnii-Tagil 'Uralvagonzavod' with 95 per cent.[111] The quality of the electronic and radio goods of MEP, MPSS and MRP is frequently criticized, with some of the harshest comments directed at their television sets.[112]

At times, even well-known, leading enterprises of the defence industry experience periods of difficulty in maintaining high-quality production. During the last two years this has applied to both 'Izhmash' and the Leningrad 'Kirov factory' association; in the latter case it led to the replacement of the general director.[113] Similarly, enterprises in Kiev and Khar'kov under Minobshchemash, in the past praised for the good design and quality of their television sets, have recently come under sharp criticism for the poor quality of their colour models. In 1983, 15 per cent of the colour televisions produced by enterprises of the ministry had to be returned for repair.[114] It is possible that at certain times the demands of the basic, military production divert the attention of management away from the civilian activity.

There is no doubt that the defence industry ministries make a sizeable contribution to civilian engineering and consumer goods exports, and the evidence suggests that this contribution has been growing over time. Successful export goods include the 'Biryusa' refrigerator (sold in many Western countries), the 'YuMZ' tractor (sold abroad as the 'Belarus-611'; the latest 'YuMZ-6AM' is to be assembled in India and Bolivia), the cameras of MOP and MAP (almost 30 per cent of production is exported) and the 'Raketa' vacuum cleaners of MAP.[115] Other products exported include bicycles (Minmash), motorcycles (MOP), consumer radio and television sets, lasers (MEP and MOP), industrial and scientific optical equipment (MOP), heavy tractors (900 exported in 1980) and, of course, the basic products such as civilian aircraft (2,000, 1971–80), ships (1,000 plus, 1976–80), hydrofoils (150 'Raketa' and 'Kometa' to 1984, one-fifth of the number built), and computers (700 plus to 1981).[116]

On the basis of the available evidence, a tenable hypothesis would be that the quality and technological level of the civilian products of the defence industry are closely related to the quality standards and rates of product renewal characteristic of the basic products of the ministries and enterprises concerned. At the same time, it would appear that the average level of quality of the civilian products of the defence sector is somewhat higher than the average level characteristic of the rest of Soviet industry.

SOME ECONOMIC ISSUES

At individual enterprises of the defence industry the scale of civilian production is substantial and runs into tens of millions of rubles per year. As indicated by accounts in the press of the work of successful producers, this civilian activity can be highly profitable and must be a source of sizeable additions to enterprise bonus funds. For one of the largest civilian producers, 'Izhmash', total profit in the five years, 1971–5, exceeded 1,000 million rubles.[117] Unfortunately, the economic costs and benefits are rarely discussed in the literature. One emigré account of the television manufacture of the Moscow Radio Factory indicates that it was taken very seriously by the management because it represented a major source of income, helped to cover the plant's expenses, and provided a source of bonuses for workers and technical and managerial staff. The television production was organized on a mass basis with a high level of mechanization. The payment of bonuses to workers of the whole enterprise was dependent on 100 per cent fulfilment of the monthly plan for televisions – a shortfall of even a single set was sufficient to deprive all employees of their monthly bonus.[118] According to Filatov, director of 'Svetlana' in Leningrad, a leading association of the electronics industry, advantages derived from consumer goods production include a greater stability of enterprise performance indicators, the possibility of achieving higher rates of growth of output and better use of production equipment. Again, one would expect these gains to be reflected in larger bonus funds.[119]

The extent to which defence industry enterprises can charge their overheads to civilian activities is not clear. Where information is available on the prices of equivalent goods produced in both the defence and civilian sectors, there is no discernable pattern of those for the former being higher. This could mean that enterprises of the defence industry have an interest in organizing their civilian production in such a way as to minimize costs through economies of scale and high levels of mechanization. To the extent that this can be achieved, it may be easier to meet the cost targets specified in military contracts by transferring overhead costs to the civilian activity. It is interesting that for a number of civilian products the largest volume producers are found in the defence sector, for example: railway wagons, motorcycles, refrigerators, bicycles and electric razors, and there appears to be strong pressure for continual expansion.[120] Some enterprises have also devoted much attention to the mechanization and automation of their civilian production, and recently this has involved the use of robots, sometimes on a substantial scale. At the Krasnoyarsk machine-building works, for example, more than 125 robots and manipulators are employed; one of the pioneers of large-scale robot use, the Kovrov Degtyarev works of MOP (making the 'Voskhod' motorcycle) now has 160.[121] This interest in robots and, more recently, flexible manufacturing systems, could signal another benefit of civilian production: in circumstances of growing labour shortage it could provide a 'captive' reserve of labour, potentially transferable to the basic, military-related production.

Against these benefits of civilian production must be set some likely disadvantages. The civilian activity may introduce an undesirable degree of uncertainty, especially production for the consumer market, and bring additional economic difficulties, including supply problems. As emerged clearly in 1970 when the Party was trying to increase the involvement of the defence industry in the production of equipment for agriculture, the civilian activity must not be allowed to interfere in any way with the fulfilment of plans for the basic production.[122] In most cases the defence industry enterprises make products for which the head ministries are located in the civilian economy. They are therefore heavily dependent on these ministries for assessments of future demand, and liable to suffer sudden changes in their production plans. This applies with particular force to the manufacture of consumer goods.[123] And in so far as the civilian ministries are under constant pressure to raise the level of specialization of their consumer goods production, especially Minlegpishchemash and the electrical engineering ministry, they are increasingly reluctant to allow defence industry enterprises to organize new production of items for which they have responsibility. At the same time, official policy, periodically reaffirmed through Party and government decrees, is for as many defence enterprises as possible to manufacture consumer goods. In the face of these uncertainties it may be preferable for enterprises to obtain permission to make industrial goods or agricultural equipment for which demand is more stable and predictable. It is interesting that the only defence industry branch with major head ministry responsibilities, Minpromsvyaz', has been very active in creating a system of demand analysis and market research, and the ministry has received high-level commendation for the way in which it is fulfilling its 'head' role.[124]

In some Western accounts of the Soviet defence industry it is taken for granted that the ministries and enterprises regard civilian production as a nuisance, and assumed that it is undertaken reluctantly under Party and government pressure. There are grounds for doubting this interpretation. As indicated above, there are certainly negative factors, but there may be benefits over and above the financial advantages already identified. The high levels of secrecy surrounding military production in the Soviet Union mean that weapons-producing enterprises gain very little publicity and their directors, managers and workers can expect to receive meagre open public recognition of their contribution. Successful civilian production can change this situation. Such enterprises as 'Izhmash', the Kirov factory, and the Krasnoyarsk works of the missile industry receive regular media attention and the names of their directors and senior managerial and technical personnel often appear in the press. This publicity and open approval of their work may bring benefits in terms of career advancement and the standing of the enterprises in relation to city and regional Party organizations and, possibly, their own ministries. Analysis of the representation of defence industry enterprise directors on the Party Central Committee and the USSR Supreme Soviet could well reveal that many come from plants with well-developed civilian activities.

CONTRIBUTION TO THE ECONOMY

The contribution to the economy of the civilian production of the defence industry can be measured in terms of the output of its diverse products. But this alone is not adequate. Above all it fails to capture the significance of this civilian activity as a mediating link between the defence industry and the rest of the economy. Western writings on the Soviet defence industry have tended to portray it as a separate, distinct sector walled off from the civilian economy by almost impenetrable barriers of secrecy. With this image has been associated the view that transfers of technology from the defence industry are almost non-existent.[125] In the author's opinion, reinforced by the present research, these traditional perceptions require modification: the boundaries between the sectors are more fluid and the transfers more extensive than is generally believed.

There are a number of ways in which the civilian activities of the defence industry may contribute to the improvement of the performance of the Soviet economy as a whole. In so far as enterprises of the industry produce high quality, exportable goods they are frequently presented as examples of good practice to be emulated by the specialized civilian ministries, providing a lever for raising the technological standards of the latter. It is possible that civilian organizations resent these comparisons, believing that their own inferior performance is at least in part a consequence of the privileged access to quality resources enjoyed by the defence industry. Nevertheless there must be production methods, managerial and organizational practices, and standards of technological discipline that can be successfully transferred without substantial resource implications.

A number of channels for transferring experience can be identified. Representatives of defence industry enterprises contribute to mainstream industrial and technical publications. Articles by specialists from 'Izhmash' and other MOP enterprises, for example, regularly appear in the motor industry's journal, *Avtomobil'naya Promyshlennost'*. Exhibitions displaying the achievements in production technology and management of leading defence industry enterprises are frequently staged at the VDNKh national exhibition centre in Moscow and elsewhere. Recent exhibitors have included 'Izhmash', the Nizhnii-Tagil 'Uralvagonzavod' and the Vyatsko-Polyansk and Kurgan machine-building factories (all MOP?), all known for their civilian activities: indeed, it is possible that they are able to exhibit in this way precisely because of their civilian role.[126] Articles in the press providing accounts of positive experience often refer to defence industry enterprises producing civilian goods, although the reader is usually left unaware of their ministerial affiliations. In the development of some of the significant new technologies these 'dual-purpose' defence industry enterprises are playing an active role, serving as transmission belts for advanced practice. Thus in the Gor'kii region, to take one of many examples, the 'lead' factories for robotics include 'Krasnoe Sormovo' (MSP – washing machines and submarines), the television works (MRP) and the aviation factory (MAP – pleasure boats).[127] In Leningrad the leading centres for the creation of flexible manufacturing systems include the Kirov factory, the 'Kalinin factory' association (MM – electric razors and tape recorders), 'Znamya Oktyabrya' (MSP – irrigators), and 'LOMO' (MOP – optical equipment).[128] Again, it may be precisely because these enterprises have developed facilities for the manufacture of civilian products that they are particularly well-suited to serve as 'demonstration centres' for new technologies, to the benefit of both the military and civilian sectors. This topic requires further investigation, but the evidence does suggest that the civilian side of the defence sector does contribute more than its products to the economy as a whole.

CONVERSION – A POSSIBLE STRATEGY?

Space precludes a detailed consideration of the problem of the potential convertibility of defence industry enterprises to civilian production. In the past some Western writers argued, rather simplistically, that social ownership of the means of production and central planning provide sufficient conditions for a rapid and smooth transition to civilian production in the event of this becoming possible. More recently, Seymour Melman and others have argued that, notwithstanding these favourable circumstances, the privileged conditions in which the defence sector has traditionally operated could present substantial obstacles to attempts to convert. The personnel of the enterprises and organizations of the industry would fear the loss of their higher rates of pay and other benefits, and would be reluctant to work without privileged access to resources.[129] In the face of such fears, it is argued, there could be reluctance on the part of the authorities to embark on conversion on any substantial scale. This argument is not without force, but the present study of the existing civilian activity of the defence industry points to a possible solution. One of the major problems of the Soviet economy is that of expanding the volume of engineering exports, in particular exports to the West. Given the success of some defence enterprises in producing quality goods which find markets abroad, if conversion became a practical possibility some enterprises could be wholly adapted to the manufacture of export goods, retaining their present privileges. Some Soviet foreign trade specialists have already raised the possibility of creating special 'export' enterprises: conversion would provide a feasible solution to the problem. In the course of time, the privileged conditions could be gradually removed, putting the former defence industry enterprises on the same footing as the more advanced enterprises of the civilian sector.

CONCLUSION

This preliminary review of the civilian activities of the Soviet defence industry indicates that the defence sector does make a substantial contribution to the Soviet economy. During the post-war years the Party and government leadership has frequently turned to the defence industry for assistance in solving priority tasks which the civilian economy alone has been unable to fulfil. To some extent this civilian involvement may serve to provide additional legitimation for the resources devoted to weapons procurement by the Soviet state. There has been a tendency for the civilian production to become more specialized so that it now represents a permanent feature of Soviet economic life, going beyond the simple use of reserve capacity. The civilian activity also provides a channel for the transfer of technology and managerial practices from the defence sector to the civilian economy and, possibly, vice versa. Many questions remain unanswered, but this preliminary study does at least indicate that the topic is researchable.

3
Soviet Microprocessors and Microcomputers

PAUL SNELL

The microprocessor heralded a new era in the West in the use of computers by society at large. The attraction of these flexible, yet cheap, control devices was obvious, and since their introduction in 1971 their availability and application have flourished. The extent of their growth has indeed been revolutionary.

What of the Soviet Union? The Soviet response to this new technology was at first sluggish, perhaps reflecting an inability to produce these devices indigenously. But by the late 1970s Soviet industry set about rectifying this, and in a very short time the USSR had made a successful bid to improve its home industry, thus creating a capability to manufacture complicated integrated circuits and to ensure substantial production of a wide range of devices.

The microprocessor in itself is not a real revolution. Its appearance could have been predicted by the steady increase in the level of miniaturization of integrated circuits (ICs) during the 1960s. The early ICs contained only a few transistors on a single 'chip' of silicon – this was the age of Small-Scale Integration (SSI). By the second half of the 1960s, several tens of transistors could be fitted on to one chip and this was called Medium-Scale Integration (MSI). The end of the 1960s saw the appearance of Large-Scale Integration (LSI) with several hundreds to a few thousand transistors on a chip smaller than a finger nail.[1] It was at this stage that circuit complexity allowed the production of a simple programmable processor on a few chips, and in 1971 the first 'microprocessor' was made available – the Intel MCS-4 series.[2] It can be argued, therefore, that the microprocessor does not represent a technological revolution, but a technological progression. However, it is the *application* of the microprocessor which truly revolutionizes society. For the first time the manufacturer has available a cheap, reliable means for the control or automation of the production process. Before the microprocessor, production control was either very expensive or inflexible, using either expensive minicomputers or fixed hardware. Now there is an 'intelligent' programmable device which, by simply changing the programme (software), can control various pieces of equipment, making all the necessary adjustments and it can even record any required measurements. The cheapness and reliability allow the automation equipment to run in a distributed (decentralized) manner with the possibility of communication links with the centre for overall control. It is in this sense, coupled with the real availability to small and large customers alike, that the term 'revolutionary' is best understood.[3]

The microprocessor is a resource-saving device. Its proper use can lead to more efficient production. As the debate in the West has emphasized, the technology is a labour-saving one: hence the charge made against the trade unions that they put job protection before technological innovation. The automation of industrial manufacturing is now possible on a hitherto unprecedented scale. This can lead to rises in labour productivity, more reliable and higher quality goods, and improvements in safety and working conditions.[4] It is not hard to see how attractive microprocessor technology would be to the cumbersome Soviet industrial 'machine'. A major problem for the Soviet Union continues to be the general labour shortage. A rapid and systematic adoption of microelectronic technology would surely reduce this. Reductions in energy and raw material wastage are, of course, other very important considerations.[5] Microcomputers can lead to more efficient running of enterprises by allowing access to production information and a more flexible control of such facets as work-in-progress. They can be used very effectively in stock control and for controlling automatic warehouses.

The scope of such improvements is by no means clear. The relative infancy of this new technology has led to much being said about savings and benefits, but few empirical studies have been conducted to compare the relative benefits of adoption and non-adoption of microelectronics in manufacturing. Nevertheless, it is clear that the emphasis on microelectronics application in the national economy has come about after serious thought and weighing up of the pros and cons. Soviet planners have seen the need for the most rapid and widespread adoption of this technology throughout the economy.[6]

Earlier studies of Soviet technological innovation have revealed certain systemic obstacles to the diffusion of new technologies throughout the economy.[7] Such studies have shown that, in general, the Soviet system is less favourable to the introduction and diffusion of new production processes and products than is Western capitalism. There are exceptions, the chief one being in the defence sector, but historically the innovative ability of the Soviet economy has been shown to be inadequate. The 'youthfulness' of microprocessor technology makes studies of the type cited extremely tentative – a longer period of time would be desirable before any firm conclusions could be drawn. Meanwhile, we have to do our best with the available material.

Going beyond the traditional, systemic (and very important) obstacles to innovation in general, it is desirable to concentrate on specific problems which the diffusion of microelectronics may face or produce. Although the releasing of labour resources is not the acute problem for the Soviet Union that it is for the West, it is nevertheless very desirable that this freed labour be retrained in order that the labour savings are not squandered. The need for computer literacy is also of paramount importance. The Soviet Union shows signs of having shortages of trained computer engineers, particularly with programming skills, hence the lucrative offers made to Western computer programmers for working in the Soviet Union in 1984.[8] Therefore, a concerted effort at re-educating the workforce and producing the correct mix of vocational students in order to provide the necessary skills is a major requirement for the effective use of the technology.

Another problem may be the incompatibility of certain production processes with the available hardware. In order to use microprocessors effectively there is the need for suitable peripheral devices and transducers for controlling processes. (A transducer is used to convert some physical change (e.g. a temperature drop) into an electronic signal understandable to the control equipment – or vice-versa. Some British firms have said that they are unable to use microprocessor technology owing to the lack of availability of suitable transducers.[9])

As suggested in this brief introduction to the benefits and problems of microelectronics, the Soviet Union would seem to stand to benefit from the introduction of this new technology into the industrial economy. Above all, microprocessors could provide a solution to the inefficient operation of production enterprises. In the West the need for this technology came about for economic reasons – the need for individual enterprises and corporations to be competitive. As this is not the case in the Soviet economy one might argue that the Soviet Union could get by without introducing the new technology. However, the overall inefficiency of production has led to many spending restraints on the Government. The best way the Soviet Union could relieve these restraints is to increase efficiency in order to provide 'more for less'. If this is not the case, then the Government may soon have to choose between raising living standards *or* raising defence spending – a choice few in power would like to make. Coupled to this is the need to remain militarily 'competitive' with the West, which probably has overriding importance in the minds of the planners. The last few years have seen a marked effort by the Soviet Union to apply this technology throughout the national economy – there can be little doubt that the Soviet authorities have made their choice.

SOVIET MICROPROCESSORS

The contradictions found in Western reports concerning Soviet microprocessors are indicative of the sparse information available to the 'public at large'. According to various Western sources, the Soviet Union lags behind the West by two to ten years,[10] copies older American devices,[11] and has nothing more advanced than an 8-bit microprocessor.[12] These claims tend to show only one part of the picture, however, and can be misleading.

By 1982 the Soviet Union claimed to be producing 15 families of microprocessors. The technologies involved cover the whole range commonly used in the West. It is claimed that the products include 4-, 8-, and 16-bit single-chip devices, and 2-, 4-, and 8-bit, bit-sliceable devices for creating longer word-lengthed processors. (Bit-slicing is where a microprocessor section is connected in parallel with similar microprocessors in order to increase the word-length for data-processing. For example, the K536IK1 is an 8-bit Arithmetic-Logic Unit (ALU) and is used in the 16-bit *Elektronika-S5* microcomputer series. This means that there are two of these chips connected in parallel for the data-processing section.) Table 3A.1 provides a list of all the chips for which technical data were found, including chips in the families which are not microprocessors. Table 3A.2 summarizes these families into a more digestible table.

It must be stressed here that this section is dealing with Soviet *claims*, and in itself cannot prove the existence of every family. However, certain families are being used in industrial or consumer equipment, suggesting their availability outside the defence sector. Nor do these claims prove the large-scale production of all the families, although, once more, some sets are thought to be readily available on a fairly large scale. What the Soviet Union seems to have done is to have developed a broad and useful microprocessor base. It has managed to cover the whole range of operation specifications for use in industrial, domestic, and most probably military applications. It has low power devices which, in the West at least, are cheap and easily used. There are the mid-range 16-bit devices for use in microcomputers for various applications where resolution and good speed are necessary. And they have the faster bipolar bit-slice devices which can be strung together to create more powerful 16-, and 32-bit machines.

There are families which are analogous to American devices: the K580 series is certainly a copy of the Intel MCS-80 series, which includes the ubiquitous Intel 8080 microprocessor;[13] the K589 series is equivalent to the Intel 3000 series;[14] and the K1810 series seems to be the same as the Intel 8086 microprocessor – it is used in the SM 1810 microcomputer which is related to the SM 1800 microcomputer. The latter uses the K580 series (Intel MCS-80), to which the Intel 8086 is related.[15] The Soviet Union also claims to possess a copy of the Intel 8086.[16] It would thus appear that the Soviet Union has drawn heavily on the work of the Intel Corporation. In addition, the K1804 bit-slice series bears a resemblance to the Advanced Micro Devices 2900 series, a popular bit-slice device family.

It should also be noted, however, that there are devices which do not have a Western counterpart: the K587 and K588 series, for example, are said to be patented in the West;[17] the K1801 series is an independently developed Soviet 16-bit single-chip microcomputer;[18] the K583 series, which uses Injection Integrated Logic (I^2L) technology, which has been noted in the West for its potential military application,[19] is again apparently uniquely Soviet. Because of the lack of information available at the time of writing, the other series must, for now, carry large question marks as to whether or not they are uniquely Soviet, or copies from Western devices. We might surmise that alongside the copying of Western devices, the Soviet Union has developed its own microprocessor base which is intended to be compatible with existing larger Soviet systems. This is explained by Balashov and Puzankov (whose names have been linked with the development of the Soviet K589 series,[20] which is based on the Intel 3000 series) in the following way: 'The following principal directions of development of indigenous [Soviet] microprocessor technology can be identified: the reproduction and development of the better foreign devices; the creation of original indigenous microprocessor chip sets; and the design of a Unified System of Microprocessors (ESMP).'[21]

Along with the work of the Soviet Union, the other CMEA countries are also developing their own microelectronic bases. Most important is the East German *kombinat* 'Mikroelektronika', which produces several series of microprocessors. These include the U880 which is pin-compatible with the popular Western microprocessor the Z80;[22] and the U830–K1883 series which is jointly designed and produced by the GDR and the USSR.[23] This is an 8-bit NMOS bit-sliceable series which is used in the SM minicomputer family.[24] Czechoslovakia is known to produce its own K589 series, which it calls the MH3000 series (reflecting more honestly its similarity to the Intel 3000 series.)[25] Bulgaria, which produces disc drives for microsystems, also claims to produce its own microprocessors. Little information is available, but they include the SM600 series which employs NMOS technology and is used in the IZOT 100 ZS microcomputer.[26] This microprocessor is also said to be equivalent to the Motorola 6800.[27]

The information in this section has shown what the Soviet literature claims is produced in the Soviet Union. This does not mean these devices are necessarily in large-scale production. However, they have been developed, and this in itself is a positive step for the Soviet electronics industry. Later it will be argued that certain devices have, beyond reasonable doubt, been produced on a large scale. These claims put in doubt some of the views generally held by Western authors.

The most noticeable omission from the Soviet Union's microprocessor base is a 32-bit microprocessor family.[28] Apart from this, the information provided suggests a very broad-based microprocessor range, covering the majority of the operational specifications required for use in most industrial applications.

SOVIET MICROCOMPUTERS

In 1984 *The Economist* stated: 'The Soviet Union is known to have produced several microcomputers based on American technology of the mid-1970s. (Only one of these is thought to be driven by a single microprocessor chip; the rest are less advanced multichip models.)'[29] This attitude is prevalent amongst Western analysts of Soviet microcomputers. Once more, it only reveals part of the truth. Since we do not know how many microprocessors, or microcomputers, are actually produced, the reference in the above quotation is really concerned with Soviet claims, and cannot refer to the scale of production. If the quotation is referring to the development of Soviet microcomputers, then once more it does not cover the whole range of Soviet claims, only the ones which have been inspected by some analysts using available Soviet literature. Other machines have largely been ignored. Technologically, the Soviet Union has developed its own devices, and they have been produced. There are certainly machines which resemble Western ones, but others have not been compared to Western machines. The Soviet Union has definitely developed, and used in microcomputers, microprocessors more advanced than the microprocessor mentioned in the above quotation, which probably refers to the Intel 8080 copy.

The present section looks at the microcomputers which the Soviet Union claims to produce. Where a machine has been compared to a Western microcomputer, this will be mentioned. As in the previous section, none of this information says anything about the production of the listed machines. This aspect will be considered in a later section. Personal Computers (PCs) are a peculiar case and hence will be dealt with separately from the microcomputers, which have a broader application.

In the microcomputer field there are four main series, with additional machines not included in these series. Some of the microprocessor families are thought to be intended for use in the ES mainframe computer series; these sets include the K583,[30] and perhaps the K1800. Included in Table 3A.3 are the details of all the machines for which technical information could be found. There are four main series of what may be loosely called 'industrial' microcomputers: the *Elektronika-60*; *Elektronika-S5*; *Elektronika-NTs*; and some of the more recent *SM-EVM* minicomputers.

(1) The *Elektronika-60 microcomputers*[31] are chiefly intended for production control and data-processing. They are 16-bit machines which can use programmes written in BASIC or FORTRAN. A variety of peripheral equipment is available. The microprocessors used in this microcomputer series are from the K581 family. The earliest microcomputer in the series was the *Elektronika-60* which had a four-chip microprocessor (16-bit); however the more recent model, the *Elektronika-60T*, uses a newer chip which combines the functions of the four earlier chips, making the later microcomputer much smaller.[32] This machine is said to be similar to the DEC LSI-11/2,[33] although this comparison is somewhat strained; it would take a careful dissection of the machine's operation system to verify or reject this supposition.

(2) The *Elektronika-S5 microcomputer series* has five or six basic models.[34] They use mainly the K536 microprocessor family, the main exception being the *Elektronika-S5-21*, which uses the K586 single-chip microprocessor. The technology used in the earlier machines is the rather slow PMOS which makes their operational characteristics quite poor. The *S5-21* uses the faster NMOS technology. They are all 16-bit machines, the earlier ones being multi-board devices and the later ones single-board. There has also been mention of a newer generation of S5 microcomputers – the *Elektronika-S5-03*, which is said to use an unspecified single-chip microcomputer.[35] These machines, which appear to be inferior compared to other Soviet models, have not been likened to any Western counterpart.

(3) The *Elektronika-NTs microcomputers*[36] are low-powered machines with a good noise immunity which makes them ideal for decentralized operation in industrial processing. They are 16-bit machines using CMOS technology. The earlier versions use the K587 series of chips, and although no single machine in this series has been identified as containing the K588 series, Soviet articles state that this microprocessor series is used in the *Elektronika-NTs* microcomputer family. The *Elektronika-NTs-80-01* contains the single-chip microcomputer K1801VE1[37] – this is an NMOS device and so the machine is faster than the earlier series. The K587 series has been patented in the West, and so these microcomputers can be regarded as being indigenously developed. This series seems to be a very respectable one, with quite good operational characteristics.

(4) *The SM-EVM series* is chiefly a minicomputer series, but recent additions are definitely microcomputers. The SM 1800 microcomputer uses the Intel 8080 look-alike, the KR580IK80A, and is hence an 8-bit machine. It is built around a very modular system using a standard interface bus called the I41 (probably the Intel 41 bus). It has been claimed that over 50 types of modules for this microcomputer are produced serially, allowing economic and reliable microprocessor systems to be built for any given application.[38] Perhaps a little more interesting is the SM 1810 microcomputer which uses the relatively new K1810BM86 microprocessor. It is a 16-bit machine, and the processor is probably a copy of the Intel 8086. Little is known about this microcomputer except that it can use the same interface bus as the SM 1800 (I41).[39] The SM minicomputer series has been compared to the Western PDP-11 minicomputers, although this may not apply to the SM 1800 and 1810 which are microcomputers.[40]

Other microcomputers are occasionally referred to, and many of them are based on the K580 series (Intel 8080), including the *Kristall-60* and *Ekspromt-80*.[41] The *Elektronika-K1-10* may also be based on this series of microprocessors, although the two have not been positively linked. These machines do not play as important a role as the four main series, owing to the simple fact that the other microcomputers are compatible with earlier machines in their relevant series and hence can be used more rapidly by industrial users familiar with these earlier models.

PERSONAL COMPUTERS

This area is a more intriguing one, with both technical and socio-political implications. These will be discussed at length later in this study. Here, we confine our attention to the characteristics of the equipment.

The *Iskra* series of computers are more like desk-top calculators. More recent additions are, however, thought to be more akin to the concept of PCs in the West. The *Iskra-250* is such a machine, and is said to be technologically advanced. Before it can be assimilated into mass-production there is a problem with the reliability of the keyboard to be overcome.[42]

The *Agat* PC is software-compatible with the Apple-II computer.[43] This is a PC which has attracted much interest in the West, owing to its similarity with the Apple – some have even called it the *Yabloko*. Internally, however, the machines are different. The *Agat* relies on the K588 series for its processor,[44] while the Apple uses the 6502 microprocessor. The *Agat* has apparently been available for hire in Moscow, but not for sale.[45] Much work is apparently being undertaken to provide back-up for this computer. One article mentions educational software for schools, its use in various sectors of the economy, and that a local network system is being developed. This has obviously been targeted as a useful PC and is set to be mass-produced, once reliability problems have been overcome. It may take two to three years to accomplish this according to E. P. Velikhov of the Academy of Sciences.[46] The same author has claimed that a

16-bit PC based around an Intel 8086 type device (probably the K1810BM86) has been produced and delivered. This is thought to refer to the *Iskra-250*, already mentioned. He has further stated that joint Soviet and Hungarian research has led to the construction of a PC in Hungary equivalent to the IBM PC, which uses the Intel 8086 16-bit microprocessor.[47]

Velikhov has raised the problem of reliability when discussing PCs, yet not in connection with 'industrial' microcomputers. The reasons for this may be organizational (reliability may vary, depending upon which ministries are involved in production) or it may be that such reliability problems in industrial machines are 'covered up'. There may also be political dimensions, which I shall discuss later in this chapter.

The evidence provided here suggests a better level of technology than claimed in the opening quotation to this section on microcomputers. There are 16-bit microcomputers whose processors are not multichip devices (e.g. the *Elektronika-S5-21* and *Elektronika-NTs-80-01*), and the range of operational characteristics of these industrial machines seems quite impressive, although proper benchmark tests would be needed to make an accurate comparison of Soviet machines with Western ones. As in the case of microprocessors, a noticeable omission is that of a 32-bit microcomputer. I have seen diagrams of 32-bit microcomputers using bit-slice devices, but these seem to remain on paper only.

THE PRODUCTION OF MICROPROCESSORS AND MICROCOMPUTERS

There are four main ministries involved in the production of electronic components and 'radioelectronic' apparatus. They are (1) the Ministry of the Electronics Industry (Minelektronprom); (2) the Ministry of the Radio Industry (Minradprom); (3) the Ministry of the Communications Equipment Industry (Minpromsvyaz; and (4) the Ministry of Instrumentation, Automation Equipment and control Systems (Minpribor). The first three ministries are part of the Soviet defence industry, which means that little is revealed about their production performance. This makes the direct quantification of Soviet production of microprocessors almost impossible. Minpribor is not part of the defence industry but seems to be more a user of electronic components than a producer. The ministry with most responsibility for microelectronics is Minelektronprom, and here most attention will be devoted to this industry. Most of the production of microelectronic components is carried out in the large production or science-production *ob"edineniya* (associations) within the branch.

The issue of production volume for microelectronic circuits is an important one, and has been at the centre of debate for a long while now. Starting from the fact that precise figures are not available, we must look closely at what is being said. Most of the information concerning microprocessor production is vague. At the Twenty-sixth Party Congress in 1981, A. I. Shokin, the Minister of the Electronics Industry, stated that: 'In the 11th Five-Year Plan the electronics industry will produce millions of microprocessors and tens of thousands of micro- and mini-computers.'[48] In one article it was claimed that the total output of Soviet microelectronic equipment – ICs, microprocessors, microcomputers, etc. – grew more than threefold in the preceding Five-Year Plan (tenth), and that in 1980 the total output exceeded 2,000 million rubles. Minpribor was cited for this information.[49] It is not known what these figures actually correspond to. They do not, for instance, correspond to the total output of Minpribor in 1980, which was 5,400 million rubles.[50] (This article was published in 1982 before the revaluation of Minpribor's output.) Neither do they correspond to the total output of computer equipment and parts, which was 4,500 million rubles for 1980.[51] The meaning of these figures may one day become clearer.

In a letter to *Ekonomicheskaya Gazeta*, in June 1982, Petr Stukolov, the Chief of the Economic Planning Main Administration of Minelektronprom, wrote that: 'Minelektronprom report that the modern state of the microelectronic technology branch has allowed the organisation of the mass-production of microprocessors and microcomputers, which are used in the construction of microprocessor systems of the most varied application.'[52] This is an important letter as it is the first time that Minelektronprom actually claimed mass-production capability, as opposed to batch or serial production. The letter then went on to mention by name the *Elecktronika-60* microcomputer, which perhaps suggests that this computer is in mass-production.

Other reports refer to the serial production of microcomputers and other such devices. One article concerning the design of an automated warehouse control system states a preference for the SM 1800 microcomputer 'as industry masters large-scale production of the latter [SM 1800]'[53] The microprocessor in this machine is the KR580IK80A (Intel 8080A). As this is a 'copy' of the Intel 8080A there is a potential market for it in the West. One US firm which imports such devices is said to have had made available from the Soviet Union over 100,000 such devices per month.[54] They are priced competitively with the cheapest such devices in the West, and it would seem that such a volume of chips available for export goes beyond being a mere political ploy by the Soviet Union to encourage the view that they are producing these devices. Thus this chip is available on a fairly large scale for export purposes, and as indicated by articles such as that cited above, is also readily available in the Soviet Union.

Philip Hanson recently remarked upon an article in *Pravda* by the minister of the machine-tool building and instrumentation industry, B. Bal'mont,[55] which expressed satisfaction with the suppliers of their microelectronic needs for the automation of machine-tools. Hanson suggested that such satisfaction with suppliers in what is a 'sellers' market' was a rare occurrence, and noted that the tone of this article was in stark contrast to that of a discussion in late 1981 on control devices for NC machines.[56] He concluded that this could be seen as circumstantial evidence of an improvement in the past two or three years.[57]

The Western concept of a 'copy' of a microprocessor threatens to oversimplify the issue, and care must be taken not to regard a Soviet 'copy' as being identical to the Western device it is based on. Closer inspection reveals such differences in the Soviet versions as chip layout and line widths. It was towards the end of 1980 when information concerning Soviet microelectronics caused a ripple in Western journals. The Control Data Corporation (CDC) released press reports on their investigations of one Soviet microprocessor, the K580IK80. Although it is equivalent to the Intel 8080, it would not be correct to say that the two chips are identical. CDC's report suggested there were differences in their structures. The investigator was told to examine the structure of the K580IK80 – a destructive examination, and consequently the device was never tested electrically.[58] However, the availability of the K580IK80 chip to the US market, as discussed above, suggests that the chips are interchangeable. In an article in *Electronics*, John Posa observed that 'the copy of the Intel 8080 was most likely laid out independently with the architectural building blocks identical to those in the US version.'[59] A few other journals wrote about the CDC report, some coming to different conclusions, but the general view of these articles was that this device, being similar to the Intel 8080A, was equivalent to the US technological level of 1977, when the Intel 8080A was released.[60] The Posa article was more detailed than the others and it questioned whether or not the K580IK80 represented the best level of Soviet technology in 1980, with CDC claiming that the chip may have been in production as early as 1977.

In summary then, although the K580IK80 is equivalent to the Intel 8080A, it is not internally identical. The Soviet engineers have studied the internal layout of the Intel 8080A and have relaid it out to suit their own production equipment. The general level of technology compared favourably with Western technology, and even more so if the K580IK80 were not 'new' in 1980.

APPLICATION OF MICROPROCESSOR TECHNOLOGY IN THE SOVIET ECONOMY

As has already been indicated, some people working with Soviet microelectronics have voiced their approval of the indigenous developments in this field. The question remains as to how well these developments are being translated into actual applications in the economy. It is not desirable here to discuss individual examples of microprocessor application; if the reader is interested then several industrial journals contain articles on this topic at regular intervals.[61]

Reading Soviet literature, one finds that the 'air of confidence' concerning their microprocessor base seems to emerge around 1979, a good deal later than in the West. But this confidence surrounding the technological accomplishments was usually accompanied by the obligatory 'howevers' when discussing the introduction of the technology into the economy. For example, V. M. Proleiko, the Chief of the Scientific Main Administration of Minelektronprom, wrote that:

Despite the fact that domestic microprocessor technology developed only recently, today we already have a series of microprocessors, microcomputers and systems, which testify to the successful and rapid development of that technology. . . . The introduction of microprocessors and microcomputers into the national economy takes place nowadays at a slower rate. . . .[62]

In outlining the problems facing the introduction of the technology, he singled out such issues as the deficiencies in the organizational structure for the dissemination of information concerning microelectronics, and the need for peripheral equipment. Such 'pessimism' was not universal. At a symposium on the application of microprocessors and microcomputers, held in Budapest in 1979, V. Ya. Kuznetsov listed several successful applications of microprocessors in the economy. However, some enterprises did not live up to expectations. The 'Soyuzpromavtomatika' *ob"edinenie* was singled out for criticism: despite a co-ordinated plan of work it had not produced a single piece of equipment based on microprocessor or microcomputer technology.[63]

The Soviet Union is only too well aware of the possible advantages of using the new technology. A first deputy chairman of the USSR State Committee for Science and Technology (GKNT), D. Zhimerin wrote:

According to preliminary estimates, the savings from the implementation of the programme [the 1982–90 programme for the use of microprocessor technology in the economy] during the period up to 1990 will be about 5,000 million rubles. The savings will be achieved by increasing labour productivity and by the improvement of product quality and the expansion of functional control. A decrease in losses and a reduction of energy and material expenditures are of no less importance.[64]

The CMEA member countries are at present working within the framework of a general agreement for the development of microprocessor systems. The 'socialist division of labour' is being implemented to meet the requirements of the member countries for peripheral equipment, software, new microcircuits and microsystems themselves.[65] The signs are that there has been a lag between the development and actual introduction of microprocessor technology; however, a concerted effort is certainly underway and the only true test of Soviet success will be time.

The political arena

The development of Soviet microelectronic technology and microcomputers brings with it some very important political repercussions. These fall, broadly speaking, into two main categories which, at some risk of oversimplification, can be labelled the 'International' and 'Domestic' aspects. The first concerns the relations between the West and the Soviet Union, and the effects of the technological embargo, while the second concerns the desirability of 'information technology' in a 'controlled' society.

Before the revision of the COCOM lists of strategic technology in July 1984,[66] there was no real discrimination between types of technology in the microelectronics field. The equation 'microelectronics = military' was used unproblematically by the Pentagon. This attitude reduced the credibility of the arguments forwarded by those whose concern it was to slow down the growth of Soviet military technology by identifying the essential areas of dependence. The military applications of a Sinclair Spectrum are probably insignificant, but the fact that the German Democratic Republic has been producing a copy of the microprocessor used in the Sinclair Spectrum,[67] plus the availability of what are superior Soviet devices, made such sanctions impractical and perhaps unwarranted. The July revisions, therefore, represented a positive step towards the attainment of the goals set out by the COCOM lists of strategic technologies – namely to slow down the growth of Soviet military technology. The COCOM lists were never intended as a general embargo, and these revisions have done much to correct this situation.

It may be true that the level of present-day Soviet microelectronic technology was achieved by the acquisition of Western process technology, legally or illegally; I am not in any real position to judge this. However, Soviet manufacturing equipment for large scale integration [LSI] and very large scale integration [VLSI] is in existence, and one such production line was exported to Hungary in 1984. The line will produce 120,000 silicon wafers annually, and the 'Mikroelektronika' institute-factory which acquired it hoped to produce two million ICs in 1984.[68]

Another product of the embargo has been the growth of a consensus in Soviet circles, that in trying to strangle Soviet microelectronics (and more generally high technology), the West, and in particular the USA, has forced the Soviet Union to develop its own industry. Some go as far as to say that the Soviet Union has actually benefited from sanctions.[69] Again, this point of view can be argued for and against – 'If they can manage without us then why did they import in the first place?', or 'Once they had the basic technology from the West, internal development became more of an option.' The reality of the situation is far from clear. The Soviet optimism can be backed by some impressive developments, but the Soviet Union still seeks to import certain types of technologies. No doubt more will be heard of these arguments in the future.

There is evidence to suggest that the production of personal computers in the Soviet Union has been delayed,[70] with Soviet writers pointing towards reliability problems as the cause. In the West it is suggested that the Soviet authorities have been trying to prevent the introduction of decentralized information systems of the type available in the West.[71] The question is whether or not the Soviet authorities welcome the introduction of PCs with network communications links. The challenge they face is that of losing control of information at the lowest tiers of society. PCs *could* be used to store *samizdat* (underground) information and creative writing and this could be stored on flexi-discs and passed around dissenting circles; if printers are available then such information could be made into 'hard copies' for more general distribution; and of course, communication networks can work both ways and protected information may be obtained by Soviet 'hackers'. All of these suggestions would imply that the Soviet authorities are facing a dilemma of sorts as to whether or not the adoption of decentralized information technology is at all desirable. Certainly, most of the Soviet effort has been towards industrial microelectronics with the successes already discussed in this study. The attitude of the people working in this field has been optimistic. Yet, when it comes to PCs, the writers are very quick to blame reliability problems which prevent their mass-production. This could be put down to the fact that Minelektronprom, the most advanced of the four ministries involved in electronics, is not directly involved in the production of PCs.[72] This points towards an uneven distribution of manufacturing technology even in the four main producer/user ministries of electronic components and equipment.

The situation is not clear, and it may be that the Soviet Union is going to be faced with a unique situation – the adoption of microelectronics to the fullest degree in industry, but the restriction of the availability of this technology to the public at large. The *Agat* PC is said to be available for hire in Moscow,[73] and perhaps this is going to be the direction PC usage in the USSR will take. Printers may be subject to the same regulations as photocopiers and other forms of printing equipment, and to obtain a hard copy of a programme the user would have to take the disc to a special office. This will not prevent discs from being handed from user to user, but if these users are then regulated by the hiring process then it starts to become apparent how the Soviet authorities could maintain some sort of control. Another very interesting aspect of this technology is the data networks. The USSR is now said to be using commercially available Western data bases.[74] Could it be long before a Western child playing at home could actually break into Soviet information bases? Even US defense computers are not immune from the prying of computer 'hackers'.

Although by no means definite, there is an apparent reluctance on the part of the authorities to allow the PC to become an everyday utensil of Soviet society. At the same time, it is perhaps pertinent to remind ourselves that it may 'only' be that the Soviet manufacturers of PCs are faced with reliability problems for the range of PCs now under production in the USSR. This latter claim is in some ways supported by recent moves by the USSR to buy PC manufacturing equipment from the West, following the revisions of the COCOM lists of strategic technologies. The talks are thought to have concerned the wholesale buying of factories to produce complete PCs, which use a microprocessor the Soviet Union claims to produce. The 'innovation' in this case may be the obtaining of reliable Western keyboards and other such hardware support.[75]

DEVELOPMENT TRENDS

The earlier series of Soviet microprocessors have not remained unchanged over the years. By 1982 several devices had been updated and improved. Previously, all the devices were in ceramic packages, now many of them are available in plastic packages.[76] Ceramic packages in the West are normally reserved for military applications owing to the ruggedness of these devices; plastic cases tend to be used in commercial equipment.[77] If this is the case in the Soviet Union it may also indicate a larger volume of production, making these devices available to 'non-priority' customers. These newer plastic-packaged devices also tend to have improved operational characteristics (except for ruggedness) over the earlier devices.[78] Some of the families have additional chips added to their number.[79] These changes point towards the familiar Soviet method of improving their equipment in progressive steps, as has been done in other sectors of the economy. Alongside this, completely new chip sets have been introduced, showing another aspect of the development of Soviet microprocessor technology.[80]

VLSI vs bit-slicing

A trend in Soviet technology has been towards developing bit-slice microprocessors. In the West the emphasis is on VLSI technology, with ever higher packing densities on the chip. VLSI technology is more advanced than the simpler bit-slice, yet bit-slicing can offer higher productivity (speed) rates than VLSI.[81] The bit-slice devices can also be more flexible in application owing to their microprogrammable structure. VLSI takes up less room and is more reliable, owing to the reduction in external connections – the cause of most failures in electronics goods. If, however, the failure is an 'on-chip' one, then bit-slicing is cheaper as only one part of the processor will need replacing rather than the whole thing, as would be the case for a VLSI device.

As suggested above, it is not a simple matter of 'ours is better than theirs'. Both trends have their respective merits and this must be borne in mind; after all, some Western manufacturers do produce bit-slice devices. One thing to note is that the 'annual world sales of microprocessors' in *Electronics* in 1982 predicted that in 1982 sales of MOS microprocessors (which includes VLSI devices) would be $251 million and that sales of bit-sliced devices would be $33 million.[82] In 1983 the same report showed that in 1982 the actual sales of MOS devices reached $173.6 million, and the bit-slice sales reached $43 million.[83] This, tentatively, suggests that the bit-slice market is growing in popularity in the West, while, perhaps not too surprisingly, the predictions for MOS devices have been over-optimistic.

This does not mean that the Soviet Union is not developing a VLSI 32-bit device; in all likelihood it will. Articles on Soviet microelectronics often mention 32-bit microprocessors and plans to develop such devices. It *could* be that they have already developed such a device, as information on new developments is very slow to seep out to the West. And of course, some of the Soviet microprocessors can be regarded as VLSI devices – the K1801VE1 being an obvious example.

The present CMEA programme for co-operation in the development of microprocessor application in the economy has been running since 1982 and lasts till 1990. The programme tries to be all-encompassing, including development, production and application of sufficient hardware and software for the needs of the fraternal socialist countries.[84] The need for specialization of production is stressed in the programme, obviously to ensure better economies of scale for each producing enterprise.[85] The requirement for improvement in Computer-Aided Design for microelectronic circuitry is emphasized in order to accelerate the development and series production of promising microprocessor components.[86] The main task is seen as sufficient co-ordination or research and production within the CMEA member countries so as to avoid wasteful repetition of work, and to meet the demand for equipment within the member countries.

CONCLUSIONS

In general the Soviet Union seems to introduce certain 'types' of microprocessor (e.g. a 16-bit single-chip microprocessor) some two to four years behind the West. The main trend of the USSR is for the fullest use of bit-slice technology, as opposed to the West's drive towards ever higher levels of integration. There are also some examples of Soviet VLSI design.

Whatever the comparative qualities of Soviet and Western devices, the evidence points towards the possibility that the USSR has sufficiently developed its technology to be able to supply industry with devices of the required specifications and quantity for use in production processes. This does not mean it can produce enough for microelectronic devices in individual consumer goods – a better knowledge of production level would be required to make that judgement, but for use in manufacturing processes there is enough evidence to support the above claim. The proper use of this technology would ensure large savings in the economy, although two of the most wasteful sectors of the economy – agriculture and construction – will not benefit to the same extent as other sectors, owing to the limited usefulness of microelectronics in these fields. In other sectors we could expect to see signs of improving productivity and better quality and quantity of products. For example, food processing could benefit significantly from the introduction of the technology. According to one British businessman the Soviet Union believes it loses 25 per cent of food owing to poor processing and packaging; Britain is currently engaged in a joint programme with the USSR in this very area.

This chapter has, it is hoped, provided a general insight into Soviet achievements in the area of microelectronics. There is room for more detailed analysis of Soviet developments, even if it is only to compare rigorously Western and Soviet devices for their comparative merits. There is also more scope for further studies in such areas as hardware and software support for the devices discussed in this study. Without adequate support from these two areas the microprocessors would be all but useless. The microprocessor was taken as only one indicator of Soviet microelectronic technology, and so our conclusions cannot be definitive.

This study shows that Soviet achievements have hitherto been underestimated in the 'popular' journals in the West. On the other hand, there is still an observable gap between Western developments and Soviet articles found in journals. This gap need not be unbridgeable, however. It is conceivable that future Soviet developments could bring them up to the US level of development, but this will depend on indigenous Soviet work rather than depending on copying Western designs. Production volume and mix remain unknown quantities, and the information provided here can be no substitute for hard figures, which will probably not be forthcoming. Finally, though, it should be stressed that it is not so important who is at the front edge of technological development. More important is how well that technology is being exploited at the industrial and services level in order to provide a more economic functioning of the respective systems. History suggests that this will provide the biggest hurdle for 'revolutionary' technology in the Soviet system.

APPENDIX TO CHAPTER 3

Table 3A.1 Soviet microprocessor series

Family	Chip	Function	Technology	Wordlength (bits)	Speed. Operation cycle (microsecs)	Power consumption (milliwatts)
K536	IK1	Arithmetic Logic Unit (ALU)	pMOS	8	10–30	70
	IK2	Microprogram unit	pMOS	8	10–30	
	IK3	Input/Output control	pMOS	8	10–30	
	IK4	Input/Output adapter	pMOS	8	10–30	
	IK5	Timer	pMOS	8	10–30	
	IK6	Voltage converter	pMOS	12	10–30	
	IR1	Buffer	pMOS	—	10–30	
	IK7	Channel selector	pMOS	—	10–30	200
	IK8	Microprogram unit	pMOS	—	10–30	70
	IK9	Arithmetic Logic Unit	pMOS	—	10–30	70
	UI1	Bus driver	pMOS	—	8	500
	UI2	Bus with memory	pMOS	—	8	500
	GG1	Control signal generator	pMOS	—	10–20	1,000
	Iv1	Keyboard coder	pMOS	—	10–30	200

Family	Chip	Function	Technology	Wordlength (bits)	Speed. Operation cycle (microsecs)	Power consumption (milliwatts)
K580	IK80	Microprocessor	nMOS	8	2	750
	IK51	Sequential adapter	nMOS	8	—	300
	IK55	Parallel adapter	nMOS	8	—	300
	IK57	Direct Memory Addressing	nMOS	8	—	500
	IK22	DRAM refresher	nMOS	—	—	—
	IK24	Control signal generator	nMOS	—	—	—
	IK53	Interval timer	nMOS	16n	—	—
	IK59	Interrupt control	nMOS	8	—	—
KR580*	VG75	Video control	nMOS	8,16	—	—
K581	IK1	Register ALU	nMOS	16	0.4	900
	IK2	Operation fulfilment control	nMOS	—	0.4	900
	RU1	System command ROM (Read only memory)	nMOS	(512×22)	0.4	160
	RU2 } RU3 }	Operation fulfilment ROM	nMOS	(512×22)	0.4	160
KR581	VE1	Combines the above 4 chips	nMOS	16	(Clock frequency) 2.5–5.3 MHz	na
	RU4	DRAM	nMOS	(16K)	—	—
	VA1	Asynchronous transceiver	nMOS	8	0.4	—
K582	IK1 } IK2 }	Microprocessor	I^{2}L	4n	1.75	725

* All the K580 chips are ceramic devices; they are available in plastic cases as well, this series being called the KR580 series, which has this single additional chip.

Table 3A.1 continued

Family	Chip	Function	Technology	Wordlength (bits)	Speed. Operation cycle (microsecs)	Power consumption (milliwatts)
K583	IK3	Microprocessor	I^{2}L	8n	1	
	KP2	Transceiver without memory	I^{2}L	4n	—	500
	KP3	Transceiver with memory	I^{2}L	4n	—	na
	KhL1	Universal switch	I^{2}L	8	0.1	330
K584	IK1 (VU1)	Microprocessor	I^{2}L	4n	2.0	130
KR584	IK1A IK1B IK1V	Microprocessor	I^{2}L	4n	2.0	na
K584	UM1	Microprogram unit	I^{2}L	16/14	0.5	na
	UM2	Status control	I^{2}L	16	0.5	na
	KP1	Bus transceiver	I^{2}L	8	0.1	na
K586	IK1	Microprocessor	nMOS	16	(Clock frequency) 0.25–2 MHz	1,000
	IK2	Input/Output unit	nMOS	8	(Clock frequency) 0.25–2 MHz	700
	RU1	Static RAM	nMOS	(254×4)	2 MHz	160
	RE1	ROM	nMOS	(1K×16)	2 MHz	240

Table 3A.1 continued

Family	Chip	Function	Technology	Wordlength (bits)	Speed. Operation cycle (microsecs)	Power consumption (milliwatts)
K587	IK1	Data exchange	CMOS	8n	1	50
	IK2	ALU	CMOS	4n	2	50
	IK3	Arithmetic extender	CMOS	8n	5–7	50
	RP1	Memory Management	CMOS	(64)	4	50
KR587	All above 4 chips available in plastic bodies					
K588	IK2 (VS1)	ALU	CMOS	16n	1.2–1.8	1
	IK3 (VU1)	Memory management	CMOS	(100)	1.5	1
	VR1	Arithmetic extender	CMOS	8n	5	1
KR588	VS2	ALU	CMOS	16	na	na
	VU2	Memory management	CMOS	(150)	na	na
	VG1	Systems controller	CMOS	—	1 MHz	
	IR1	Multimode buffer	CMOS	8	—	
	VA1	Bus transceiver	CMOS	8	—	

Table 3A.1 continued

Family	Chip	Function	Technology	Wordlength (bits)	Speed. Operation cycle (microsecs)	Power consumption (milliwatts)
K589	IK01	Microprogram control	TTL Sch. D	—	0.2	400
	IK02	Processor element	TTL Sch. D	2n	0.15	850
	IK03	High-speed carry	TTL Sch. D	8	0.02	400
	IR12	Buffer	TTL Sch. D	8	0.08	450
	IK14	Priority interrupt	TTL Sch. D	—	0.08	450
	AP16	Bus driver	TTL Sch. D	4	0.02	450
	AP26	Bus driver inverted	TTL Sch. D	4	0.02	450
	Kh14	Counter unit	TTL Sch. D	4	0.02	450
	RE4	Microprogramm-able memory. 1K	TTL Sch. D	4	0.04	500
K1800	VS1	ALU	ECL	4n	36 MHz	na
	VB2	Synchronization circuit	ECL	—	36 MHz	na
	VT3	RAM (random access memory)	ECL	4n	36 MHz	na
	VR8	Programmable driver	ECL	16n	36 MHz	na
K1801	VE1	Single-chip microcomputer	nMOS	16	8 MHz	na
	VM1	Microprocessor	nMOS	16	5 MHz	na
	VP1	Matrix chip	nMOS	—	—	na

Table 3A.1 continued

Family	Chip	Function	Technology	Wordlength (bits)	Speed. Operation cycle (microsecs)	Power consumption (milliwatts)
K1802	VS1	ALU	TTL Sch. D	8n	8 MHz	na
	IR1	General purpose registers	TTL Sch. D	(16×4)n	10 MHz	na
	VR1	Arithmetic expander	TTL Sch. D	16n	8 MHz	na
	VV1	Data exchange	TTL Sch. D	(4×4)n	10 MHz	na
	VR2	Serial multiplier	TTL Sch. D	(8×8)n	8 MHz	na
	VV2	Interface adapter	TTL Sch. D		10 MHz	na
K1803	VE1	Single-chip microcomputer	CMOS	4	1 MHz	na
KR1804	VS1	Microprocessor section	TTL Sch. D	4n	8 MHz	na
	VU1 } VU2	Microcommand series control	TTL Sch. D		8 MHz	na
	VR1	High-speed carry	TTL Sch. D		8 MHz	na
	VU3	Next address selector	TTL Sch. D	32×8	8 MHz	na

Table 3A.1 continued

Family	Chip	Function	Technology	Wordlength (bits)	Speed. Operation cycle (microsecs)	Power consumption (milliwatts)
	IR4	Parallel register D-type	TTL Sch. D	4n	8 MHz	na
	VS3	Arithmetic microprocessor	TTL Sch. D	4n	8 MHz	na
	VR2	Status and shift register	TTL Sch. D	4n	8 MHz	na
	VU4	Microcommand sequencer	TTL Sch. D		8 MHz	na
K1810	BM86	Microprocessor	nMOS	16	10 MHz	na

Source: A. A. Vasenkov, *Mikroprotsessornye BIS i Mikro-EVM*, M. 1980, pp. 14–17. A. A. Vasenkov, *Mikroprotsessornye komplekty integral'nykh skhem*, M. 1982, pp. 179–83. E. P. Balashov and D. V. Puzankov, *Mikroprotsessory i mikroprotsessornye sistemy*, M. 1981, pp. 292–4.

Table 3A.2 Summary of Soviet microprocessor families

Family	No. of Chips in Set	Technology	Word length	Notes
K536	14	pMOS	8	Bit-sliceable, used in *Elektronika-60* and *Elektronika-S5* microcomputers.
K580	10	nMOS	8	Based on the Intel MCS-8 series.
K581	5	nMOS	16	Used in *Elektronika-60* microcomputer.
K582	2	I^2L	4n	Bit-sliceable.
K583	4–12	I^2L	8n	Bit-sliceable.
K584	7	I^2L	4n	Bit-sliceable.
K586	4–9	nMOS	16	Used in *Elektronika-S5-21* microcomputer.
K587	4	CMOS	8n	Bit-sliceable. Patented in the West. Used in the *Elektronika-NTs* microcomputers.
K588	8	CMOS	16	Used in *Agat* PC.
K589	9	TTL Schottky	2/4/8	Bit-sliceable. Based on Intel 3000 series.
K1800	4	ECL	4n	Bit-sliceable.
K1801	3	nMOS	16	Set includes a single-chip microcomputer.
K1802	6	TTL Schottky	8/16n	
K1803	?	CMOS	4	Single-chip microcomputer, no data available.
K1804	9	TTL Schottky	4n	Bit-sliceable. Possibly based on AMD2900 series.
K1810	?	nMOS	16	Based on Intel 8086 series.

Source: Summary of Table 3A.1.

Table 3A.3 Soviet microcomputers

Microcomputer	Speed 1000 ops/sec	ROM/RAM size	Chip Series used	Word length
Elektronika-60	130–250	1K×22/ 4K×16	K536/581	16
Elektronika-60T	250–300	1K×22/ 4K×16	KR581	16
Elektronika-S5-01	10	2K/3K (×16)	K536	16
Elektronika-S5-02	10	2K/10K	K536	16
Elektronika-S5-11	10	1K/128	K536	16
Elektronika-S5-12	10	2K/128	K536	16
Elektronika-S5-21	180	2K/256	K586	16
Elektronika-NTs-03	—	—	—	16
Elektronika-NTs-03T	100	8K	K587	16
Elektronika-NTs-03D	100	16K	K587	16
Elektronika-NTs-31	130	8K/32K	(K587?)	16
Elektronika-NTs-04T	200	32K		16
Elektronika-NTs-05	1300			16
Elektronika-NTs-80-01	550/250	16K/16K	K1803	16/32
Elektronika-NTs-80-01D	500	56K/8K	K1803	16/32
SM-1800			K580	8
SM-1810			K1810	16
SM-50, 51, 52, 53, 54	Up to 500	Various up to 64K	K580	8
Ekspromt-80	300	4K/8K	K580	8
Agat PC	300	16K/32, 64, 128K	K588	8
Iskra-250			K1810(?)	16

Sources: A. A. Vasenkov, *Mikroprotsessornye BIS i mikro-EVM*, M. 1980, pp. 161–213; E. P. Balashov and D. V. Puzankov, *Mikroprotsessory i mikroprotsessornye sistemy*, M. 1981, pp. 307–16; A. V. Giglyavyi, *Mikro-EVM SM-1800: Arkhitektura, Programmirovanie, Primenenie*, 1984; N. B. Mozhaeva, *Pribory i sistemy upravleniya*, 1983, No. 12, pp. 37–40; *Nauka i zhizn'*, 1984, No. 10, p. 90; A. A. Vasenkov, 'Razvitie mikroprotsessorov i mikro-EVM semeistva "Elektronika-NTs" na osnove kompleksno-tselevykh programm', *Elektronnaya Promyshlennost'*, 1979, No. 11–12, pp. 13–17.

4
Soviet Biotechnology: the Case of Single Cell Protein

ANTHONY RIMMINGTON

Single cell protein (or SCP) is the name given to the protein-rich microbial cells which are produced when bacteria, algae, fungi or yeasts are grown on a variety of feedstocks. These latter have in the past included petroleum, natural gas, methanol, ethanol, almost every variety of agricultural waste, gases in mineshafts, peat, timber waste – even sunlight. The SCP so produced is then fed to people either directly, by addition to traditional foodstuffs, or indirectly and much less efficiently, via animals.

In the Soviet Union, which now has one of the largest microbiological industries in the world,[1] attention has been focused on SCP as a means of solving a crucial problem: the provision of an adequate supply of desperately needed protein to the livestock sector of agriculture. Given the fact that SCP production in the West has now run into very serious obstacles, it is of considerable interest to observe how this branch of industrial biotechnology has fared within the framework of the Soviet economic system.

If it were given priority over other sectors, there would be ample raw materials for such a programme; the Soviet Union possesses the world's largest forested area and natural gas reserves, and has vast stocks of petroleum and peat available. The fact that SCP production relies on such feedstocks and is not subject to the vagaries of the weather, unlike other, alternative sources of protein such as soya, will undoubtedly have influenced Soviet planners in their selection of the microbiological industry for the task of filling the protein gap.

There are also strategic reasons for undertaking such a programme. Dependence on other countries and especially America to meet the requirements of the livestock sector is undoubtedly viewed as a very unsatisfactory state of affairs. The Soviet planners may also have been aware of the successful utilization of SCP for feeding both animals and people in Germany during two world wars.[2] Perhaps, more ominously, it adds to the Soviet microbiological warfare capability, if only indirectly, by increasing the technical expertise of those involved in the production of microbes. The reported release of anthrax spores from one microbiological establishment will have served as an adequate warning to Western governments.[3]

One of the major reasons for the failure of SCP ventures in the West has been political and environmental opposition. For example, BP had to abandon their 100,000 tonnes per annum plant at Sarroch in Sardinia (after incurring development costs of £100 million) having perhaps 'fallen foul of environmental authorities out to prove their toughness in the wake of such disasters as Seveso'.[4] The powerful soya importers lobby may also have exerted pressure on the Italian authorities. The absence of such phenomena in the USSR must endow Soviet biotechnology with definite advantages over its Western counterparts. But just how well has the microbiological industry fared over the past two decades? In order to answer this question we must examine the history and development of Soviet biotechnology, which is the subject of the next section.

HISTORY

In the Soviet Union the first large-scale production of SCP occurred during the Second World War.[5] There was at this time a chronic shortage of foods rich in protein and, as part of the solution to this problem, a scheme was worked out for the production of SCP from wood-shavings (in November 1941 by V. I. Sharkova and co-workers in Leningrad).[6] By the spring of 1942 several plants were producing SCP according to this method and 'yeast steaks' and 'yeast milk' were then prepared from it for consumption in the restaurants of Leningrad.[7] Later, in Moscow in the period 1942–3, and in the larger towns of the Urals and Siberia in the period 1943–4, 'dozens' of small yeast-producing plants were brought on-stream.[8]

Thus, for the first time in the Soviet Union, microbial biomass was perceived as constituting a valuable source of protein, and the military significance of SCP production became apparent, for example, food could be produced directly in the cities without the need to import grain from war-affected areas. In this respect it is interesting that Germany managed to replace as much as 60 per cent of pre-war period imported foodstuffs by SCP during the First World War,[9] and again, Nazi Germany was producing some 15,000 tonnes per annum in 1936.[10]

In the Soviet Union, the microbiological industry was neglected during the first post-war Five-Year Plans. It consisted of several hydrolysis plants producing ethanol, feed yeasts, carbonic acids and furfural from wood-pulp hydrolysates, agricultural wastes and sulfite liquors.[11] But by the beginning of the 1960s the further construction of such factories had been discontinued. Ethanol could now be produced more economically by oil refineries and 'moreover, the main consumer of industrial ethanol, the synthetic rubber industry, began to transfer to a new technology – obtaining divinyl on the basis of direct synthesis from butane.'[12]

It appears that at this time there was a change in emphasis within the hydrolysis industry towards the large-scale manufacture of SCP. At it turned out, a number of circumstances favoured this switch in production. These include the facts that the micro-organisms already in use, *Saccharomyces* and *Candida*, were later shown to be ideal for the safe, rapid production of SCP, and that 'it was a relatively easy job to convert the anaerobic alcohol processes to aerobic SCP processes.'[13] As well as the conversion of existing hydrolysis factories, several specialized hydrolysis plants (with capacities of 10–14,000 tonnes) were constructed for the production of SCP from hydrolysate sugars.[14]

But during the 1960s other important developments were taking place which would also be very significant for the long-term future of the Soviet microbiological industry. At this time, economic planners within the Council of Ministers appear to have viewed the industrial production of SCP as being the best solution to the growing protein gap in animal feed production. As a result, the microbiological industry was given greater priority for its tasks and the Main Administration of the Microbiological Industry (*Glavnoe upravlenie mikrobiologicheskoi promyshlennosti*) was formed to oversee two distinct branches of SCP production; the first was based on the pre-existing All-Union Hydrolysis Agency (*Gidrolizprom*) and its associated research institute, the All-Union Scientific Research Institute for Hydrolysis (*VNIIgidroliz*) in Leningrad; the second, consisting of the All-Union Scientific Research Institute for Protein Synthesis (*VNIIsintezbelok*) situated in Moscow, was created in 1963 with the task of obtaining microbial protein from petroleum.[15]

The decision to opt for a second, alternative path of development, probably reflected an ongoing debate within the Council of Ministers at this time. Undoubtedly, the facts that the crude oil found in the huge Volga–Ural fields in 1960 had a high n-alkane (waxy) content (making it harder to pump and handle, and thus decreasing its value) and that micro-organisms utilize just this very fraction, must have made protein production (and accompanying dewaxing) by this method, appear very attractive.[16]

The first investigations into the production of SCP from petroleum fractions were carried out in 1960 at the Institute for the Biochemistry and Physiology of Micro-organisms at Pushchino.[17] Following on from the work of Dworkin and Forster in 1954–6, three scientists at the Institute, G. K. Skryabin, A. B. Losinov and V. K. Eroshin were awarded the Lenin Prize for their development of an n-alkane process which utilized the yeast *Candida guilliermondii*.[18]

By 1964, an experimental plant for the production of feed yeast from petroleum (with an output of 1,500 tonnes per annum) had been started up in Krasnodar.[19] Tests showed, however, that the use of impure gas-oil substrate led to toxicological problems. A switch to purified liquid alkanes as substrate in a second experimental plant at Ufa (capacity 12,000 tonnes) removed this obstacle. A wide range of biological tests demonstrated the safety of using the plant's product (given the name 'BVK' or protein-vitamin concentrate) in animal feedstuffs. However, one specialist at least is wary of 'optimistic assertions' concerning the side-effects of such biomass.[20] And, while A. A. Pokrovsky, the head of the Institute for Nutrition in Moscow, was willing to approve the use of SCP for livestock, he did add, 'We do not share the views of some scientists eager to speed up SCP application for human consumption since there may still be considerable risks, but we feel that systematic scientific research should be conducted in this area.'[21] This may suggest that there is some debate in the USSR concerning the use of SCP, at least for food purposes.

The composition of BVK in terms of protein, amino acids and vitamins seems to compare quite favourably with traditional sources of animal feedstuffs. One major problem, though, has been the low content of methionine in the finished product.[22] It appears that the Soviet Union has turned to Western technology for a solution, as in 1980 Tekhmashimport of Moscow signed a contract with Speichim of France worth 875 million French francs for the supply by the latter of a factory producing methionine (using processes developed by Rhone-Poulenc and others).[23]

After the successful trials, construction of large-scale enterprises (utilizing n-paraffins as feed stocks) took place at Ufa (100,000 tonnes per annum) in 1968, Gorki (100,000 tonnes per annum, later expanded to 200,000 in 1970), and Kirishi (100,000 tonnes per annum) in 1972.[24] Other n-paraffin plants also appear to have been built at Angarsk, Krasnodar, Kstovo, Novocherkassk, Novogor'ki, Polotsk and Svetloyarsk (see table 4.1). And at least three more enterprises at Kremenchug,[25] Novopolotsk[26] and Mozyr[27] are nearing completion and may even have come on stream (capacities of 120,000, 120,000 and 300,000 tonnes per annum respectively). The latter plant (at Mozyr) was in fact constructed under the terms of a CMEA agreement signed in 1979 by representatives of the GDR, Cuba, Poland, and the USSR and Czechoslovakia, according to which the Russians have agreed to supply some of the SCP produced to the co-signatories in exchange for help given during construction.[28]

But this ambitious programme has not been completely problem-free. At least up until 1976, in the Soviet Union, liquid paraffins and the accompanying deparaffinizates were produced on carbamide deparaffinization apparatuses.[29] However, a major drawback was that the use of the impure n-paraffins produced by this method as a SCP feedstock meant that the resulting biomass had to undergo extractive purification to meet the necessary quality standards. Even with such purification, the SCP being produced may still have been contaminated with carcinogenic polycyclic aromatic compounds derived from the n-alkanes.[30] Table 4.2 shows the analysis of the SCP produced by BVK plants in the mid-1960s (the only period for which data are available). If one compares these figures with the recommended standards for single cell biomass as published by the International Union for Pure and Applied Chemistry (IUPAC) in 1979, which require a share of hydrocarbons of less than 0.5 per cent, then it is clear that the SCP being produced in 1966, at least, would not have satisfied these requirements (in only one case was the standard met).

Hence attention within the USSR now became focused on the use of highly purified liquid paraffins (with n-paraffin content not less than 99 per cent and aromatic hydrocarbons not more than 0.01 per cent) as raw material for the production of SCP.[31] Commercial methods for producing pure n-paraffins by molecular sieve or urea adduction techniques 'have been in use for many years, providing material for the paint, solvent and detergent industries'.[32] Despite this, the fact that the Soviets had to approach Union Carbide[33] to enquire about a molecular sieve process, suggests that their own indigenously developed 'Ufa process'[34] (the separation of n-paraffins from isoparaffins, aromatics and cyclo aliphates by formation of solid urea adducts) was unsatisfactory.

At present it would seem that a gradual switch to pure n-paraffins is taking place within the industry (e.g. changeover to this feedstock in the Kirishi plant in 1977)[35] and this should lead to an improvement on previous SCP yields of about 1 kg of SCP for each 1 kg of pure n-paraffins consumed.[36]

Table 4.1 Production of SCP from n-paraffins

Plant Location	Feedstock	Products	Status	Capacity (Tonnes per year)
Angarsk[a] (Irkutsk Oblast': RSFSR)	n-paraffin[a]	SCP[a]	On-stream but construction is continuing.	?
Gorki[b] (Mogilev-skaya oblast': Belorussian SSR)	n-paraffin[b]	SCP[b]	Construction began 1970; new capacity to be added.[b]	200,000[b]
Kirishi[c] (Leningrad oblast': RSFSR)	n-paraffin wood	SCP[c]	Construction began around 1973–4. 'Became fully operative for two weeks in December 1977'. Hydrolysis of wood uses 80 and 160m^3 fermenters made of titanium alloy.[c] Protein content of 59 per cent aimed for.[d]	100,000[b,e]
Krasnodar[f] (Krasnodarskii krai; RSFSR)	n-paraffin sunflower husks rice husks tanning waste molasse treacle (Waste from sugar refinings)[f] ?	SCP[f] Ethanol[f] Furfural[f] Carbonic acid[f] Xylitol[f] Fuel bricks[f] etc.	One of the microbiological industries largest plants.[f]	8,000[f]
Kremenchug[g] (Poltavskaya oblast': Ukranian SSR)	n-paraffin	SCP[g]	Under construction[g]	120,000[g]
Kstovo[h] (Gor'kovskaya oblast': RSFSR)	n-paraffin	SCP[h]	On-stream[c]	?
Mozyr[i] (Gormel'skaya oblast': Belorussian SSR)	n-paraffin	SCP[i]	Not on-stream?	300,000[i]
Novogor'ki[j]	n-paraffin	SCP[j]	?	?
Novocherkassk[k] Rostovskaya oblast': RSFSR	n-paraffin	SCP[k]	?	?

cont.

Table 4.1 (continued)

Plant Location	Feedstock	Products	Status	Capacity (Tonnes per year)
Novopolotsk[l] (Vitebskaya oblast': Belorussian SSR)	n-paraffin[l]	SCP[l]	First put into operation in March 1978 – probably not achieving full capacity.[l] A second production line was put into operation in 1981.[m]	120,000[n]
Polotsk[o] (Viteb-skaya oblast': Belorussian SSR)	n-paraffin[o]	SCP[o]	?	?
Svetloyarsk[p] (Volgograd oblast': RSFSR)	n-paraffin[p]	SCP[p]	Largest enterprise for the production of this product in the country.[p] In 1980 was the first plant to exceed an output of 100,000 t of BVK a year.[p] Target of 240,000 t of BVK a year.[p]	>100,000[p]
UFA[q]	n-paraffin[q]	SCP[q]	According to Humphrey, construction began in 1968[b] but first product not obtained until 1975.[q] Fourteen fermenters.[q]	100,000[q]

[a] *Ekonomicheskaya Gazeta*, 1983, No. 51, p. 1.
[b] A.E. Humphrey 'Single cell protein: a case study in the utilization of technology in the Soviet system' in J. R. Thomas and U. M. Kruse-Vaucienne (eds), *Soviet Science and Technology: Domestic and Foreign Perspectives*, George Washington University, 1977.
[c] *Mikrobiologicheskaya Promyshlennost' (Nauchno-tekhnicheskii Referativnyi Sbornik)*, 6 (180), 1981, p. 29.
[d] *Mikrobiologicheskaya Promyshlennost' (Nauchno-tekhnicheskii Referativnyi Sbornik)*, 7 (149), 1977, p. 5.
[e] *Ekonomicheskaya Gazeta*, 1979, No. 12.
[f] *Mikrobiologicheskaya Promyshlennost' (Nauchno-tekhnicheskii Referativnyi Sbornik)*, 3 (173), 1980, p. 30.
[g] *Pravda*, 2 June 1980.
[h] *Mikrobiologicheskaya Promyshlennost' (Nauchno-tekhnicheskii Referativnyi Sbornik)*, 3 (177), 1981, p. 25.
[i] *Ekonomicheskoe Sotrudnichestvo Stran-Chlenov SEV*, 1984, No. 1, p. 38.
[j] *Ekonomicheskaya Gazeta*, 1975, No. 32, p. 2.
[k] *Mikrobiologicheskii Sintez (inform. materialy)*, 1966, No. 11, p. 11.
[l] *Mikrobiologicheskaya Promyshlennost' (Nauchno-tekhnicheskii Referativnyi Sbornik)*, 2 (188), 1983, p. 24.
[m] *Mikrobiologicheskaya Promyshlennost' (Nauchno-tekhnicheskii Referativnyi Sbornik)*, 5 (170), 1981, p. 1.
[n] *BBC Summary of World Broadcasts (SWB)*: SU/W1123/A19; *SWB* SU/W1161/A16.
[o] *Mikrobiologicheskaya Promyshlennost' (Nauchno-tekhnicheskii Referativnyi Sbornik)*, 3 (177), 1981, p. 19.
[p] *Mikrobiologicheskaya Promyshlennost' (Nauchno-tekhnicheskii Referativnyi Sbornik)*, 6 (186), 1982, p. 10.
[q] *Gidroliznoe Proizvodstvo (Nauchno-tekhnicheski Referativnyi Sbornik)*, 5 (137), 1982, p. 10.

Table 4.2 Analysis of SCP produced by BVK (n-paraffin) plants

Plant location	Moisture	Salts	Content (per cent)		Lipids	Hydrocarbons
			Total protein	Carbohydrate		
Krasnodar (1966)	10.00	12.45	48.69		4.31	0.64
Krasnodar (1966)	8.90	9.25	50.47	12.79	3.50	0.75
Krasnodar (1966)	8.99	8.25	51.80		1.80	0.45
Tavda	9.33	7.90	50.28	22.61	3.88	1.03
Novocherkassk ('Biomeal' analysed)	9.50	7.20	46.63	19.98	2.50	0.80

Source: V. P. Aristova, L. P. Khayurova and A. V. Skryabina, 'O gigroskopicheskitch Svoistvakh sutkikh kormovykh drozhzhei poluchennykh na osnove ochishchennykh zhidkikh parafinov nefti', *Microbiologicheskii Sintez*, 1967, No. 5, p. 5.

The petrochemical industry has had major difficulties in meeting the demands of the microbiological industry for n-paraffins, as the complaints of one author from the Ufa factory testify.[37] The main reason for this is a lag in the construction and bringing on-stream of liquid paraffin plants, for example at the Novogor'ki, Novo-Ufa and Syrazan oil-refineries.[38] Thus, the commissioning of the Novopolotsk BVK factory was delayed because it could not start operation until the Novopolotsk liquid paraffin plant was completed.[39]

But the Soviet Union is now facing an even greater problem in relation to its use of n-paraffins as feed-stock: there are indications that there may be difficulties in maintaining supplies of this raw material in the future. By 1976 the microbiological industry had already become the largest consumer of liquid paraffins, accounting for 60 per cent of the total volume produced in that year.[40] Bobrovskii et al. estimated that by 1980 the proportion of petroleum with a 'high' and 'average' paraffin content in the total extracted would be decreasing, with a simultaneous sharp increase in the volume of petroleum with a 'low' paraffin content. Hence, they concluded that the potential resources of liquid paraffins will be reduced in the future.[41] Obviously this could impose severe restraints on the expansion of this branch of the industry and bring into question the wisdom of constructing huge plants such as those at Mozyr, Kremenchug and Novopolotsk, which may have been commissioned before such 'shortages' were envisaged.

Another problem encountered (and indicated in the literature) is that of environmental pollution. In 1977 there was a call for improvements in the purification installations of the Kirishi factory.[42] By 1981 these had still not been carried out and there were reports of discharges of protein gas into the atmosphere.[43] However, it would appear that the introduction of a system of 'damp purification' has now largely solved the problem in this particular factory.[44] For the microbiological industry more generally, a series of measures has now been worked out to increase the efficiency of gas purification and to reduce discharges into the atmosphere. Dust-catching installations are being installed with capacities of $1,150,000$ m^3 per hour.[45]

One of the most effective ways of reducing pollution would be to produce SCP in a granulated form.[46] At present the Soviet Union is still producing yeasts in a powdered form, although there are problems with humidity (which reaches levels of up to 10 per cent), dust and bulk-transportation of the finished product.[47] Many plants may be perfectly capable of producing yeasts in a granulated form but unfortunately the state standard (GOST) forbids this.[48] Despite complaints (even from Rychkov, the Head of the Main Administration of the Microbiological Industry) the situation still remains the same.[49] Thus, this is a clear example of a GOST holding-up scientific and technical progress. In the longer term there are plans to produce granulated SCP but even by 1990 this is expected to form only 60 per cent of total output.[50]

The health of workers in this branch of the industry is also causing concern.

A study of the health of workers at a pilot plant producing BVK and at a heavy-tonnage biochemical combine showed the major change resulting from contact with the production ingredients (the yeast and the paraffin) involved local and diffuse skin lesions with a direct ratio between severity and length of contact. Skin tests showed positive reactions of varying degree to paraffin and the finished product.[51]

What effect all this may have on the productivity and morale of workers within the industry remains unclear.

A major problem which results from work being carried out in non-sterile conditions is that of contamination with foreign microflora. Koval'skii et al. conducted a detailed analysis of SCP produced at the Krasnodar and Ufa plants to investigate this matter.[52] They carried out 216 analyses of the finished product and showed that the microflora infecting the SCP were bacterial forms and mainly sporophytic and saprophytic fungi.[53]

The total quantity of bacteria in the SCP was shown to fluctuate between 500 and 51,850 cells per gram (there was also seasonal fluctuation in the quantity of infective bacteria).[54] The composition of the sporophytic bacteria was quite stable and in the main represented by *Bacillus subtilis*. Of the rest, *B. megathernium*, *B. mesentericus* and *B. mycoides* were also regularly distinguished, and also certain other unidentified rod-shaped forms. In separate tests and in small quantities, *B. cereus* was encountered.[55]

Sporophytic, anaerobic microflora were represented by *Clostridium perfringenes* and *Cl. sporogenes*. The quantity of Clostridia in the SCP depended upon how many were present in the concluding stages of the manufacturing process. Koval'skii et al. showed that the total quantity of this group of bacteria in the finished product is from 3 to 1,150 cells per gram (see table 4.3).[56]

Non-sporophytic microflora found in the SCP included bacteria from the families *Pseudomonas, Baterium, Proteus, Sarcina,* and *Micrococcus*. The SCP becomes contaminated with these species when interruptions occur in the flow of yeast suspension (creating stagnant zones) or upon exposure to the air.[57] Another source of pollution of SCP, especially by saprophytic fungi, was reported as being through contact of the dry yeasts with fresh medium.[58]

The results suggest that the SCP being produced in the Soviet Union (for these two plants at least) does comply with the microbiological standards accepted in the West (table 4.4 shows these standards as published by IUPAC in 1979). The viable bacteria count as measured by Koval'skii was always well below 100,000 cells per gram and in nearly all cases the number of Clostridia were below 1,000 per gram. However, the quantity of *Cl. perfringenes* in the sample was never specified by these workers and may well have exceeded the IUPAC recommendations. Also, unfortunately, no information was provided as regards viable molds and streptococci. Nevertheless as far as one can tell, the microbiological industry in the USSR is currently producing SCP with a generally acceptable level of contamination as defined by Western specialists.

Yet another problem has been indicated by Humphrey, who reports that over $100,000 was spent by the Soviet Union in 1974 on a computer-coupled fermenter system constructed in the USA to optimize their SCP process. However, even as late as December 1975 the system had not yet been made operable due to a lack of 'expertise in computer programming and in instrumentation repair'. Reports such as these suggest that serious technological problems have arisen within the Soviet industry over the last two decades and that recourse to Western expertise and 'know-how' has often been the best, if

Table 4.3 Quantity of infective bacteria in SCP

Year of production	Manufacturer of BVK	Number of research tests on total content of bacteria	Total quantity of bacterial cells in 1g of yeast			Number of research tests on content of Clostridia	Total quantity of Clostridia in 1g of yeast		
			Minimum	Maximum	Average		Minimum	Maximum	Average
1970	Ufa (Experimental industrial BVK factory)	16	500	12,400	7,448	—	—	—	—
1971	Krasnodar chemical combine	13	1,000	11,500	5,722	10	3	130	65
	Ufa	58	810	51,850	7,353	54	50	1,150	389
1972	Krasnodar	15	1,000	21,100	12,750	17	90	320	190
	Ufa	20	1,300	12,000	5,965	19	350	600	452
1973	Ufa	22	2,400	40,000	10,899	22	260	1,100	830

Source: Y. u. V. Koval'skii, N. B. Gradova, V. R. Arkhipova and S. A. Konovalov, 'Mikrobiologicheskaya kharakteristika gotovogo produkta sredakh s n-parafinami', in *Mikrobiologicheskaya Promyshlennost', Referativnyi Sbornik,* 7 (127), 1975, p. 12.

Table 4.4 Microbiological standards of new or unconventional proteins for use in animal feeds (IUPAC recommendation 1979)

	Per gram
Viable bacteria count (total aerobic bacteria)	<100,000
Viable molds	<100
Enterobacteriaceae	<10
Salmonella	<1 per 50 grammes
Staph aureus	<1
Clostridia total	<1,000
Lancefield group D Streptococci	<10,000

Source: A. Einsele, 'Biomass from higher n-alkanes', in H. Dellweg (ed.), *Biotechnology*, Vol. 3, Weinheim, Basel, Verlag Chemie, 1983, p. 79.

not the only, method of overcoming them.

My previous work has tended to present the microbiological industry as being divided into two distinct branches for the production of SCP (one utilizing wood and agricultural wastes – the hydrolysis industry; the other utilizing petroleum and natural gas) with little or no overlap.[59] In the light of the information now available this view will have to be altered. For example, the plant at Kirishi (Leningrad oblast') not only produces SCP from liquid paraffins but also from wood wastes via hydrolysis (the separate technical processes were developed by *VNIIsintezbelok* and *VNIIgidroliz*; the actual design of the factory being carried out by *Giprobiosintez* of Leningrad).[60] Apparently, both the closeness of an oil-refining factory and the large quantity of timber available locally influenced the siting of this particular factory.[61] Similarly, for the Krasnodar plant, SCP and other products are being produced from both n-paraffins and agricultural wastes.[62] How far this situation is typical of all plants producing SCP from liquid n-paraffins remains to be seen. However, it would appear to be a logical development, allowing all the possible sources of raw materials for SCP within a particular area to be utilized without the need of a separate factory for each type.

What, though, of future trends within this branch of the microbiological industry? One direction which the Soviets may take is the more efficient utilization of waste substances produced by the BVK factories. At the Ufa plant, for example, a method has been developed for the processing of a secondary product, 'activated sludge' which contains up to 45 per cent protein and many microelements.[63] This product is sterilized and added to the fermenters where cultivation of the biomass is taking place, thereby reducing the expenditure of liquid paraffin. For the time being, however, this valuable product is being buried and the potential savings are not being realized.[64]

Plans are also afoot to utilize the lipid fraction of yeasts grown on liquid paraffins and a pilot plant is already under construction for the extraction and processing of lipids.[65] The latter will be used in the production of high-quality soaps (triglycerides), in lubricating oil, in the medical industry (phospholipids) and in the paint industry (oleic, linoleic acids). Biologically-active substances such as ergosterol may also be isolated from the lipid fraction.[66]

Automated production lines and more powerful fermenters are to be introduced into BVK factories.[67] In fact, there has been a trend towards an increase in the individual capacities of fermenters over the past ten years and currently up to 30–35 tonnes of protein per day can be produced.[68] However, the new ADR-76 fermenter with a design capacity of 50 tonnes per day which has been installed at *Glavmikrobioprom* factories has not attained the projected output and its design is to be altered.[69]

In tandem with the development of SCP production from petroleum, the hydrolysis industry has also made significant advances. Many of the pre-existing plants were reconstructed for SCP production and many new plants built. Various feedstocks were used, including wood-chips, sawdust, corn-cob cores, rice husks, sunflower husks, cotton hulls, bagasse and molasses waste, wood-pulp and sulphite liquor. Today, the majority of SCP plants in the USSR appear to use such non-hydrocarbon feedstocks.[70] There follows a more detailed examination of the current state of the industry, beginning with fodder yeast production in pulp and paper combines.

SCP PRODUCTION FROM SULPHITE WASTE LIQUOR (SWL) IN THE USSR

If one compares the figures for Soviet pulp production with those of the Swedish industry, then it emerges that in the USSR sulphite pulp forms a larger proportion of total pulp production.[71] Thus, whatever the reasons for this disparity (possibly the relative backwardness of the Soviet industry), it would appear that there are greater opportunities for yeast production from SWL in the USSR than in Sweden and possibly also other Western countries.

Most of the Soviet pulp and paper plants have a capacity at or below some 6,000 tonnes of SCP/year, although there seems to have been a trend towards the construction of larger yeast plants in pulp and paper combines commissioned from the mid-1960s onwards. Overall, total SCP production in the pulp and paper industry is given by one Soviet source[72] as 142,194 tonnes in 1977 and there seems to be no cause for doubting this figure, which is confirmed in an independent Western study.[73]

Criteria for measuring the quality of SCP from carbohydrate sources have been laid down by the Commission of Fermentation of IUPAC.[74] Amongst the recommendations were those regarding toxicological and nutritional tests, analytical procedures and also 'safety aspects, storage measures and physical properties'. Little information on these aspects of Soviet SCP production from SWL has been found to date, although data have been published concerning the percentage of crude protein. In 1977 this ranged from 41.6 to 51.1 per cent, with an average value of 46.9 per cent (45.4 per cent in 1975), which does not compare very favourably with SCP being produced by similar processes in the West (around 55 per cent).

One of the major problems facing Soviet biotechnologists in producing SCP from SWL is low productivity. There are two main reasons for this; first, the fermenters being used at these plants may have low oxygen-transfer capacity (Humphrey reports approaches to the Austrian firm Vogelbusch for equipment with 'excellent oxygen transfer capacity'); secondly, there is the problem of the low content of reducing substances (RS) in the basic raw materials (e.g. a total RS content of only 0.6–0.7 per cent is typical for those plants – more than half of the total number – processing a dilute sulphite-ethanol waste).[75]

The East German industry apparently faces precisely the same problems and as well as developing new fermenter systems, scientists in the GDR are planning to add materials with a high carbon concentration, to effluents which have a low RS concentration.[76] One such material being actively considered is methanol, which is fairly cheap, easily transportable and available in large quantities – probably from the Soviet Union. Trials have been carried out with various substrate pairs and the improvement in yield from such additions is reckoned to be approximately 10 per cent. Whether the Soviet Union will pursue these paths is not certain, but the close ties that exist between her and the GDR would clearly facilitate such a development (the vast Soviet methanol production capacity may also influence the planners).

As well as increasing the efficiency of existing SCP production, there are possibilities for an expansion of capacity. Sopko et al. estimate that in order for rapidly increasing demands in paper and paperboard to be met, 'in a relatively short time 15 to 18 new complexes have to be constructed'. They also predict that each of these complexes would typically be fitted with a feed yeast factory of some 50,000 tonnes per annum capacity.

Thus, if Sopko et al. are correct about the demands for pulp and paper in the USSR, and if the Soviet Union goes ahead with such a programme (as outlined above), a further 750–900,000 tonnes of SCP may be produced each year within the industry. Clearly this would represent a vast increase over current capabilities and such estimates must remain highly speculative. However, this does illustrate the potentiality of the pulp and paper industry as a means of increasing overall SCP production, and also how the prospects for such production can be intimately associated with the expansion and development of an entirely separate industry.

SCP FROM LIGNO-CELLULOSE WASTES: THE HYDROLYSIS INDUSTRY

In a far from comprehensive survey, I have identified some 41 hydrolysis plants operating in the USSR. From these data no definite conclusions can be drawn about the average size of yeast shops within hydrolysis factories, but it would appear to be the case that the development of the industry continued during the 1970s with the construction of several large-capacity plants. Whether this trend will carry on into the 1980s remains to be seen. But the fact that the sylvo-chemical industry now appears to be an economically viable alternative to production of chemicals from petroleum[77] is surely significant, and may ensure its continued expansion.

Humphrey quotes 'knowledgeable US process engineers' as feeling 'that the Soviet process is not significantly different from that developed by the US Department of Agriculture during World War II'. In the former, apart from the substrate (wood-chips, sawdust, and other sources), typical raw materials for the production of SCP are sulphuric acid, superphosphate, acqueous ammonia, potassium chloride, lime, urea, lysine, soda and fishoil or soapstock.[78]

Yeasts of the genus *Candida* are mostly used, but during production the fermenter often becomes contaminated with low-yielding strains which pass in with the water and nutrient salts.[79] Thus, such species as *Zygofabospora, Torulopsis famata, T. marxiana, T. holmii, T. aeria, Cryptococcus fesreus*, and *C. mycoderma*, 'that are widespread in hydrolysis factories'[80] give a yield of only 25–30 per cent. Currently, a conversion to sterile conditions would 'permit better control of the microbial synthesis process'[81] but this has yet to be achieved.

The All-Union Scientific Research Institute of the Hydrolysis Industry (*VNIIgidroliz*) is selecting better strains which give good results in the laboratory (e.g. 'Tul-6' with a biomass yield of 56–8 per cent; growth rate of $0.25 hr^{-1}$; crude protein content of 54 per cent).[82] But problems still remain at the factories, where the strains recommended by *VNIIgidroliz* are not becoming properly established: this points to the necessity of each enterprise developing its own strains or modifying those produced by *VNIIgidroliz*. But this does not appear to be within the capacity of factory research departments, especially when, as Semushina indicates, there is a lack of qualified personnel and equipment. This may represent a major obstacle to the achievement of efficiency within the industry.

A new trend in this branch is the production of carbohydrate feeds.[83] This involves the direct conversion of cellulosic materials such as straw and the wastes of the potato- and starch-producing industries into protein-aceous feeds (i.e. the wastes are enriched with protein to make them utilizable as feeds).

The process proceeds as follows: first, the straw (or other waste material) is partially hydrolyzed by weak solutions of sulphuric acid. Then, following neutralization, it is innoculated with yeast or fungi. The mycellium fungus of the Spicaria genus gave the best results in trials, with 48–52 per cent of protein (in the mycellium) and giving a 60 per cent yield of the RSs in the hydrolyzate. The fungus raised the protein content of the fodder by 5.8 per cent.

Production of such feeds has now begun at several plants, (for example, Kirov, Manturovo, Boksitogorsk), although their quality may not be very high.[84] Kamyenni calls for the 'production at our plants of a balanced feed for livestock, which will include in addition to carbohydrate yeast, vitamins, minerals and other elements'.

CURRENT PRODUCTION AND PROSPECTS

Figures showing the production of SCP in the USSR over the past 20 years are presented in table 4.5. It can be seen that the rate of growth is not at all constant and in one year, 1979, production actually fell (but this was a bad year for the chemical industry as a whole and almost certainly has nothing to do with specific failings within the microbiological industry).

While this almost unique, large-scale manufacture of microbial products must be admired, nevertheless the production figures, as they stand now, do represent a failure on the part of the planners to implement the necessary capacity. For the goal set by the Council of Ministers for 1980 was some two million tonnes of SCP,[85] and at the current rate of progress it may well be 1990 before that target is reached – and all this must be set against a background of a rapidly increasing demand for protein products to enrich animal feeds.

Soviet economic planners wishing to avoid the import of such protein additives may now be faced with the choice of a vast increase in capacity of what has been a fairly successful microbiological industry, or they may turn to other sources of protein (for example, soya) produced by agriculture. Perhaps, more likely, they will opt for a combination of the two approaches. But in the case of the former, how will the necessary expansion be brought about?

Many Soviet experts recognize that the available resources of petroleum (and, in particular, paraffins), given competition for their use, are insufficient for such an increase in production. The fact that 'natural gas output is growing far faster than crude petroleum output'[86] has attracted the attention of Soviet specialists to the former as a substrate for SCP production. A pilot plant for the production of SCP from natural gas has been constructed at the Svetloyarsk BVK factory in the Volgograd oblast' (RSFSR).[87] The unique technology and equipment involved in this project were developed by scientists from *VNIIsintezbelok* and the GDR.[88] Output of 'Gaprin' (the name given to the plant's product) began in December 1983. The Soviet Union hopes to use the technology developed in this pilot plant at enterprises which are to be built all over the country.[89] The eventual aim seems to be the construction of enterprises 'of unprecedented capacity (300,000 to 500,000 tonnes of output per year) in immediate proximity to gas fields or in areas of large-scale protein consumption'.[90]

Another, and perhaps more promising, method of utilizing natural gas is via methanol. Indeed, there are reports that the USSR has been negotiating with ICI for the purchase of their 'Pruteen' technology (which utilizes methanol from natural gas).[91] The story of the possible Soviet–ICI deal is in fact a rather complex one. And to understand the background, one has to look at related projects in several parts of the world. Since 1973, ICI have been receiving cheap North Sea gas (6p a therm) for methanol production under the terms of a secret agreement with the British Government.[92] This contract has now come to an end and the company has turned its attention to the Middle East and the USSR where gas might be purchased at some 3p a therm in return for the 'suitcase full of technological tricks and complete marketing networks'[93] which ICI is able to offer in exchange.

On 5 January 1984, ICI were reported as stating that 'the Arabian Gulf is just one of several areas where we are looking'.[94] Certainly the Middle East must appear very attractive – Saudi Arabia has recently built a gas-gathering complex at a cost of some £8 billion, from which they will produce methanol at world-beating prices. ICI were 'slow to spot the danger' of this new source of

Table 4.5 Production of SCP in the USSR

Year	Production of SCP (tonnes)
1960	16,700[7]
1961	23,300[5]
1962	38,100[5]
1963	57,200[5]
1964	76,000[5]
1965	98,000[1]
1966	124,000[4]
1967	158,000[4]
1968	196,000[4]
1969	227,000[4]
1970	261,000[1]
1971	314,000[3]
1972	365,000[3]
1973	446,000[3]
1974	540,000[3]
1975	674,000[1]
1976	819,000[2]
1977	904,000[2]
1978	918,000[2]
1979	862,000[2]
1980	923,000[1]
1981	1,043,000[1]
1982	1,140,000[6]
1983	1,319,000[8]

Sources:
[1] *Norodnoe khozyaistvo SSSR (Narkhoz)*, 1922–82, Moscow, 1982, p. 188.
[2] *Narkhoz*, 1981, p. 163.
[3] *Narkhoz*, 1977, p. 218.
[4] *Narkhoz*, 1976, p. 254.
[5] *Narkhoz*, 1972, p. 221.
[6] *Narkhoz*, 1982, p. 150.
[7] *Narkhoz*, 1972, p. 221.
[8] *Narkhoz*, 1983, p. 159.

competition and reportedly turned down an offer to be involved in a joint venture with SABIC (Saudi Basic Industries Corporation – the state-run construction company), and must be regretting this decision in the light of lucrative deals between Exxon, Shell, Mobil, Mitsubishi, and the Saudis.[95]

However, ICI's Pruteen technology may be of some considerable interest to the Saudis. A feasibility study of SCP production carried out in the late 1970s, showed that such production was economically viable, but the Saudis were 'not fully convinced about the suitability of these products for animal feed'[96] (reports from Europe and America talked about potential hazards and side-effects).

At present, the Saudis seem to be monitoring the results of world-wide production processes, but in 1981 'were a long way from making a decision on full-scale commercial operations'. The fact that any such project they did initiate would 'utilize locally produced methanol'[98] means that ICI could possibly include the 'ready-to-use' Pruteen process (also utilizing methanol) in any deal struck between them and the Saudis.

Very significantly, ICI has just completed a £5 million deal with Saudi Arabia for the construction of a plant (at Dammon on the Gulf Coast) to produce polyurethanes for insulating factories and other uses. This is a 'significant first step towards establishing an integrated petrochemical operation in Saudi Arabia, based on Saudi hydrocarbon feedstocks'.[98] Erlichman reports that negotiations concerning ICI's Pruteen process have started with OAPEC (Organization of Arabian Petroleum Exporting Countries).[99] These facts taken together might suggest much closer future links between ICI and the Saudis.

Another country which has both gas surpluses and a protein deficiency is Mexico, and she has recently expressed interest in the Pruteen process.[100] This could be significant as the Mexicans already have at least two SCP plants producing some 20,000 tonnes per annum of SCP from molasses and carbon dioxide.[101]

However, ICI already has very close links with the Soviet Union – a long-term trade and co-operation agreement having been signed.[102] And, 'in October, ICI took co-operation one stage further with an agreement to apply Western technology, chemicals and methods to four widely different 500-hectare plots in four separate areas of the Soviet Union [with a view to raising] yields on them to Western levels.'[103]

A number of other factors favour the conclusion of a deal by ICI with the Soviet Union for the purchase of methanol. Mr John Harvey-Jones, ICI's new Chairman, is apparently a fluent Russian speaker 'who is eager to expand the trading relationship which already exists between ICI and the Soviet bloc'.[104]

In fact, methanol plant technology has already been sold by ICI under a product buy-back deal with the Russians, and two plants were ordered in 1977, the first of which came on-stream in 1983.[105] As part of this deal, ICI were to receive one-fifth of the output and are currently receiving 'the first trickle of methanol supplies which ought to reach 300,000 tonnes a year when the agreement is in full force.'[106]

The Russians are also 'extremely eager to get their hands on ICI's world-leading Pruteen process' and this latter point seems to be the key to any agreement being successfully concluded. In Autumn 1983, a Soviet deputy Prime Minister, Mr Leonid Kostandov, visited ICI's Pruteen plant at Billingham and initiated 'high level negotiations between ICI and the Russians'. An ICI delegation planned to visit the Soviet Union in February.[107] And, on 16 May 1984, Mr John Harvey-Jones, Chairman of ICI, flew to Moscow where he had talks with the Prime Minister, Mr Nikolai Tikhonov and four deputy Prime Ministers.[108] The talks were 'expected to centre on how ICI can contribute to the Soviet agriculture production through the provision of Pruteen technology'.[109] The fact that this was the first visit to the Soviet Union by a director in ten years[110] shows how much importance is attached to this deal by the British company.

Unfortunately for ICI, their main champion and one of the chief proponents in favour of a deal to purchase the Pruteen technology, Leonid Kostandov, died in May 1984.[111] Despite this setback, the close links which ICI already have with the Soviet Union certainly favour the conclusion of an agreement, perhaps in the very near future. At a recent Symposium on Soviet science and technology, an ICI representative talked of the possibility of the Soviet Union purchasing five to ten plants (each with a 100,000 tonnes capacity). Thus the USSR now perhaps has the opportunity of adding one million tonnes per annum SCP production to its existing capacity through the import of Western technology. Possibly ICI would receive cheap supplies of methanol in exchange.

The use of ethanol as a feedstock has also been considered.[112] A plant in Ufa has undergone reconstruction and is now producing SCP from ethanol for large-scale testing in agriculture. Preliminary results suggest that the yeast so derived may be of sufficiently high quality to enable it to be used for human consumption.[113] Plans exist for the construction of a 100,000 tonnes per annum capacity pilot plant based on ethanol.[114]

There is an alternative to the use of hydrocarbons and natural gas in particular for SCP production – that is the hydrolysis industry. Expansion of this branch could be achieved through an increase in the supply of existing feedstocks or by the development of new ones. Thus the use of peat (the USSR possesses vast reserves, 210,000,000 tonnes being extracted annually) has attracted the attention of Soviet specialists.[115] Following the successful operation of a pilot plant at the 'Ezherelis' (Lithuania) peat enterprise,[116] a plant for the production of fodder yeasts and feed additives (1,500 and 7,000 tonnes respectively) from peat, is now under construction at Zilais Kalns[117] (with another at Tomsk),[118] and plans have been mooted for the construction of 98 plants to produce 1,965,000 tonnes per annum of SCP from peat.[119] There are similar ideas concerning the millions of tonnes of cotton waste produced in Uzbekistan and elsewhere.[120] But just how far this programme is capable of meeting the requirements of feed production in the Soviet Union remains to be seen.

IS BIOTECHNOLOGY FEEDING THE RUSSIANS?

Ultimately, the success or otherwise of the microbiological industry in the USSR can be measured only by its contribution to domestic livestock production – the original *raison d'être*. Indeed, the demand for microbial protein is now greater than ever before. *Ekonomicheskaya Gazeta*, for example, reports that in 1983 fodder was required for some 93.7 million cattle, 61.5 million pigs and 123 million sheep and goats. Clearly, these are huge requirements by any standards.[121]

There is also an urgent need for an improvement in the quality of feed being supplied to the livestock sector. Serious protein deficiencies exist: raw-protein levels per feed unit stand at only 85 per cent of the Soviet planners' norm, and Zahn reports that there has been no appreciable improvement in this situation over the last 15 years.[122] The overall protein deficit is also very serious, with, for example, a 1.8 million tonne protein deficiency in the Ukraine. This figure represents some 11 million tonnes of oat equivalents.[123] Very large amounts of grain could be saved if the protein deficiency were to be made up.

SCP could make a very valuable contribution to the solution of these problems. For example, Rychkov (Head of the Main Administration of the Microbiological Industry) points to the fact that the addition of one tonne of SCP feed ensures the production of an extra 0.4–0.6 tonnes of pork, or 25–30,000 eggs, or releases 5–7 tonnes of grain.[124] But despite feed yeasts having been used to enrich 20 million tonnes of animal feed in 1981, they have, according to Zahn, 'played only a minor role in the total feed-supply picture. Certainly, these commodities play a significant role in particular subsectors of the livestock economy; but, overall their contribution has been minor' (Zahn estimates this at 1.1 million tonnes of oat equivalents in 1978/9).

One of the contributory factors to this lack of impact on the part of SCP is probably inadequate quality. A Soviet source, for example, complains of feed yeasts being supplied in a 'compacted form' to the feed enterprises.[125] But the main reason must be that SCP is simply not being produced in the quantity demanded. Rychkov estimates that requirements are only being met by some 25–30 per cent. Thus, an expansion of SCP production capacity is essential if the microbiological industry is to maintain its position as one of the front runners in the protein-supply field.

Zahn also indicates that part of the solution to the overall protein deficit problem lies in the fact that there is a deficiency of lysine in current Soviet protein sources. But it does seem that in this area at least the microbiological industry is making big strides forward. Although production only began in 1970, by 1980 it had reached some 9,000 tonnes with plans for a doubling of capacity by 1985.[126]

Thus biotechnology has a long way to go before it can claim to be 'feeding the Russians'. Nevertheless, it has made significant advances over the past two decades and the growth in output of SCP will at least have helped to alleviate the protein deficiencies outlined above. Undoubtedly, the Soviet Union will view the development of this industry as an outstanding success, given the spectacular failures that have occurred elsewhere. And in the long-term, especially if North America were to face a prolonged period of poor harvests, they may indeed be proved right.

5
Soviet Product Quality, State Standards and Technical Progress*

MALCOLM R. HILL and RICHARD McKAY

During the late 1960s a leading Soviet economist, D. S. L'vov,[1] observed that 'the word "quality" figures frequently in economic literature. Dozens of different definitions of this concept exist, each of which have their own particular feature different from the others'; and he then went on to discuss the term from the philosophical, engineering, legal, economic and social points of view. Looking at the engineering aspect of quality in more detail, L'vov considered that

[this] aspect of quality is related to research into technical, quantifiable laws covering the formation and manifestation of physical, mechanical, chemical and other properties, of items of identical functional purpose. From this point of view, quality is commonly considered as the totality of properties of a product, which determine the possibility of its utilisation in service.[2]

The latter sentence of this definition is almost identical to that quoted in a British Standard published in 1972, namely 'the totality of features and characteristics of a product or service that bear on its ability to satisfy a given need'.[3] There is, therefore, a close correspondence between Soviet and Western concepts of quality, particularly in what is often referred to as the 'quantitative sense' of quality[4] (or 'quality level'), where product requirements can be expressed according to technical concepts and quantitative parameters.

L'vov's definition of quality is also similar to that frequently quoted by Western writers on this topic at about the same time, namely 'fitness for purpose'. Lockyer,[5] in his widely read general textbook on production management, described the quality of a product by that term as does another leading writer in this field, J. M. Juran.[6] The concept of 'fitness for purpose', however, requires some thought to be given to the economic aspects of product quality, which are normally more difficult to define. This difficulty in definition arises from the necessity of obtaining detailed information on manufacturing costs, which can vary with volume of production and capital investment; and customers' utilization costs, which the seller is not always in a position to know.

*The research described in this chapter has been financially supported by a grant from the Joint Committee of the Science and Engineering Research Council and the Economic and Social Research Council. We also wish to acknowledge the co-operative assistance of the library of the British Standards Institution (BSI) at Milton Keynes, during the period of this research.

Furthermore, it is even more difficult to define accurately the effect of quality on profits and profitability of the manufacturer and user; although these two indicators are considered to influence the continued existence of a Western company, and the success of a Soviet enterprise.

Thus, in the research programme discussed in this chapter, particular attention is being paid to the technical, rather than the economic, aspects of product quality. It may be argued that such an approach may provide an incomplete picture of product quality, but we considered it better to reflect usual industrial practice in this present research by focusing on technical indicators. In practice, these indicators are established at values considered to provide economic utilization of a product in the most commonly expected range of applications, although products may also be ranked into 'grades' where this is feasible.[7] This latter practice then permits the selection of a higher (and probably more expensive) grade of product to meet a more demanding application.

L'vov also discussed the legal aspect of product quality, which he considered to 'be related to the manner in which legally, or contractually, specified technical requirements are adhered to by the relevant manufacturing organisations'.[8] This, in its turn, can be related to his concepts of 'engineering quality' when 'the properties ... of functional purpose' are included in 'appropriate legal and contractual technical documentation'. These Soviet concepts of the legal aspects of product quality are probably not very different in day-to-day practical application from those encountered by Western industrial managers in their own experience in their own countries. What does differ in this field between the Soviet and Western economies, however, is the legal status of various technical documents and the associated parameters specified within them, in view of the different roles played by government bodies in the USSR and the West in the planning and control of industrial activities. This theme of the legal status of various technical documents is referred to again in more detail in a subsequent section of this chapter.

PREVIOUS STUDIES OF SOVIET PRODUCT QUALITY

The general view of Western scholars is that the quality levels of Soviet industrial products are generally lower than their Western counterparts as a result of the influence of several factors. According to Berliner,[9] the major factors can be summarized as the low levels of labour skills during the early years of Soviet industrialization, problems encountered in the allocation of supplies in a centrally-planned economy, and the continued existence of a seller's market in the Soviet Union. Since the publication of Berliner's book in 1976, there have been two further studies by Grant,[10] and Treml,[11] which also appear to substantiate the view of a generally low level of Soviet product quality in particular sectors of industry. A study by Gorlin[12] on consumer goods also reaches similar conclusions, although that publication is more concerned with product 'style' than with technical specification.

Grant's conclusions will be considered first, since his is the earliest of the studies. He pays particular attention to the Soviet machine-tool industry, classifying machine tools into two broad groups, namely 'conventional' and 'advanced' machine tools. Conventional machine tools are defined by Grant as those types that have traditionally been produced: namely, lathes, drilling, boring, grinding and milling machines, and transfer lines; 'advanced' machine tools are defined as conventional types that have been enhanced in one or more key aspects (flexibility, productivity, precision) through the application of electronics and computers. Considering conventional machine tools first, Grant concludes that 'most of the machine tools that reach Soviet standards can be assumed to be less precise than their Western counterparts. That is because Soviet accuracy requirements for precision machine tools tend to be less stringent than corresponding Western requirements.'[13] In the same paper, Grant also cites several Soviet and Western sources including those compiled by Berry[14] which claim that certain Soviet machine tools lose their initial accuracy in a shorter time than their Western counterparts, lack certain design features, and are also less reliable and durable. Grant summarizes his view of these sources by stating that 'Soviet conventional machine tools do not differ technologically from those of the developed West – in design and principle of operation they follow world-wide practice – but mainly in quality: that is in performance, durability and reliability.'[15] Finally, Grant also claims that there are major technological lags in Soviet 'advanced' machine tools, compared with their Western counterparts, caused mainly by technical deficiencies in control equipment.

Grant's conclusions on the accuracy of Soviet conventional machine tools were not supported by any associated technical data, however, but partly by the citation of a previous study carried out by Hill[16] on a sample of Soviet state standards which had been approved before 1970; although that same study went on to show that improvements in state standards were apparent after 1970. It is therefore our view that some of Grant's conclusions on Soviet standards for ex-supplier machine-tool precision, and his associated conclusions on machine-tool quality, should be treated with a certain amount of circumspection.

On the other hand, a study carried out by Hill[17] on a small sample of Soviet milling machines, drilling machines and grinding machines, purchased by a British factory in the later 1960s, would certainly support Grant's second cited conclusion relating to performance, durability and reliability. That study revealed certain shortcomings in the design and manufacture of those machines which subsequently affected their working speeds, continued accuracy, reliability, and down-times; even though the initial tolerances as specified in the state standards, and achieved in the alignment tests, were reasonably satisfactory. If several of these 'design-based' shortcomings (for example, thermal distortion of grinding-head spindle bearings) were not rectified on machinery specially tooled for use in the intensive production conditions of the high-volume industries,[18] problems of manufacturing efficiency would almost certainly occur. In addition, a sample of British machine-tool engineers with experience of selling into the Soviet market from the late 1960s to the mid-1970s, when interviewed by the author in 1978, were of the opinion that Western machine tools designed for use in the high-volume industries were more productive than their Soviet counterparts. This was due to greater Western experience in the design and use of specialized tooling and to the use of more reliable and durable assemblies, components and materials.[19] The importance of numerical control interface component performance and reliability was also raised by one engineer with experience in this field. Few of the engineers, however, raised serious doubts over the initial accuracy of Soviet machine tools.

Treml's paper[20] is an informative account of the price supplements received by Soviet enterprises to cover the costs associated with modification of their products for sale in export markets. These supplements average out at 24 per cent for 'general destination' exports and 58 per cent for 'tropical destination' exports; a Soviet source[21] cited by Treml indicates similar levels of price supplements for machine tools. Treml relates these price differences to quality differences, stating that 'export price supplements appear to be used, that is quality improvements and modifications are required for practically all machine exports, which suggests that the quality gap between domestic and world standards pertains to most machinery produced in the USSR, except possibly military products.'[22] This view also appears to be supported by the comparatively low market shares achieved by Soviet engineering exports in the Western economies[23] even after supplementary work for export has been carried out by their manufacturer.

It is my opinion, however, that additional costs are incurred by many exporters in Western countries also, as a consequence of such factors as difference in electrical equipment specification, special safety and testing procedures required in certain markets, and other technical regulations which sometimes appear to be specifically designed to protect domestic producers and to discourage the penetration of the market by imported products.[24] Consequently, similar additional costs might also be faced by Soviet enterprises producing machinery for the export market. It is Treml's view, however, that for American computers, aircraft, electronic components and machine tools, the need for modification for other markets is minimal; whilst Swedish products usually only cost about 4 per cent[25] more to adapt for export markets. The magnitude of the additional costs in the Soviet case, therefore, does lend strong support to Treml's view that the major factor influencing price differences between machine models for domestic and export markets, is the quality difference; but this still requires testing by information from more actual cases. Furthermore, it is important to note that the comparatively low market shares achieved by Soviet products may not be entirely attributable to quality levels, since there may be many marketing factors worthy of further study, such as range, accessories,[26] service, and customers' perceptions.[27]

THE CASE FOR FURTHER RESEARCH ON SOVIET PRODUCT QUALITY

The studies referred to in the previous section have provided some useful insights into the quality of articles manufactured in the Soviet economy, but few of them have carried out investigations into the quality of specific products. It is considered, therefore, that further research is necessary for the following reasons:

(1) It is difficult, for example, to assess accurately the performance of the Soviet economy and to compare it with its Western counterparts, if the quality of industrial production is omitted from economic and financial calculations. If poorer quality items are manufactured in the USSR there may be an overall decrease in industrial efficiency, whilst if items of too high a quality are manufactured, a waste of valuable resources may result. The study of aspects of quality of Soviet manufactured articles is therefore of possible use in making a more realistic assessment of Soviet growth rates, capital accumulation, and industrial efficiency for comparative purposes.

(2) It is of general interest to study the methods of product quality regulation in a centrally-planned economy where, compared with Western economies, there is limited opportunity for purchasers to transfer their buying power between different suppliers. This latter aspect may be particularly relevant in the purchase of capital goods, the quality of which can subsequently affect the quality of items produced by them.

(3) It is likely that particular Soviet import objectives have been related to the country's need for certain high-quality capital goods, which it has not yet sufficiently developed for manufacture on a wide scale, but which are required for the expansion of a priority industry. Thus, it is useful for prospective and current exporters to the USSR to be acquainted with the current quality level of similar Soviet-produced articles, and the ways and means that such parameters are maintained and modified, in order that market requirements may be more clearly defined.

(4) Since the success of industrial products in Western markets is influenced to a very great extent by non-price factors, such as quality and service, it is useful to assess the degree to which Soviet market shares in Western markets[28] for industrial products may be influenced by their quality/price/service indicators: and the extent to which Soviet manufacturers may need to modify products for export to the market economies.

(5) It is useful to explore more thoroughly the relationships between Soviet product quality and Soviet industrial innovations, to determine comparative performance characteristics for novel products, and those products at a mature stage of their life-cycle.

This chapter, however, is more concerned with quantitative aspects of product quality management, and specific ways in which product quality can be comparatively assessed as a means of providing information to answer some of the questions raised in the previous section. Hence, we shall discuss product standards as convenient sources for comparative quality assessment.

STANDARDIZATION IN THE SOVIET ECONOMY

Engineering standard specifications are documents approved by a recognized authority at the relevant level, which specify rationalized dimensional parameters and defined quality characteristics of industrial articles. Standardization is used in all industrial countries as a means of promoting production specialization through variety reduction, and ensuring that the quality of manufactured articles is to an acceptable level.

Standards can be published by many different bodies in market and mixed economies.[29] Individual Western companies may, for example, publish standards for components and materials drawn up by the company's Standards Department for their own internal use, in order to obtain a more intensive use of a restricted variety of these items and hence reduce production costs. Furthermore, such standards may form the basis of purchasing orders to ensure the quality of incoming raw materials and components. The majority of Western companies also produce what they refer to as 'standard items', which are frequently those made in batches for serving customers from stock, or those items which have been manufactured previously for which relevant technical documentation exists. In such cases, it is usually the company's Sales Department which is the 'recognized authority', on which products can be considered as 'standard'. A further set of standards used in Western companies are those approved by a national standardizing body (e.g. British Standards Institution) in a Western country. It is unusual, however, for these Western national standards to have legal status in their own right, and they are usually applied voluntarily by those companies to which their requirements apply. They may however, form the basis of a purchasing contract and thereby assume legal status, or they may be incorporated into a country's legislation covering such items as health and safety.

In the USSR, individual enterprises, like their Western counterparts use standards for the purpose of attempting to reduce manufacturing costs and particularly to help guarantee the quality of sources of supply. These documents are referred to as 'Enterprise Standards' (*standarty predpriyatiya*), and may form the basis of a purchasing contract. Similarly, industrial ministries may also publish standards for use by those factories responsible to them. The national standards used in the USSR, (referred to as 'state standards' or GOSTy (abbreviation for *gosudarstvennye standarty*)), however, have legal status in their own right,[30] since they are considered to be a means of ensuring that the state is provided with a supply of products of adequate quality. They automatically form a legal framework for purchasing contracts between Soviet enterprises, as well as carrying penalties for non-observance. Hence, it is important for 'industrial' and 'enterprise' standards to conform with state standards for those products specified by the latter.

The current framework for publishing state standards in the USSR has emerged since 1965, following the re-establishment of the ministerial system of industrial management,[31] and the passing of two resolutions by the Council of Ministers of the USSR to strengthen the role played by standardization in Soviet economic development.[32] Many of the features of this curent framework are not new, however, since standards have been published for national use by Soviet industry for some 60 years, the first central organization responsible for standardization being set up by government decree in 1925. This organization, named the All-Union Committee of Standardization, was responsible for the approval of industrial standards which had legal status throughout the national economy. The Committee underwent several organizational changes, until in 1954 it was made responsible to the Council of Ministers of the USSR, (the highest government body in the Soviet Union), and its title changed to the Committee of Standards, Measures and Measuring Instruments under the Council of Ministers of the USSR.[33] From the beginning of 1971, its title was modified to the State Committee of Standards of the Council of Ministers of the USSR, and its Chairman elevated to membership of the Council of Ministers of the USSR. In spite of the Committee's apparent improvement in status, however, there appears to be little change in its administrative framework formalized between 1966 and 1970, although its powers of arbitrary inspection of the quality of produce of industrial enterprises have been increased.[34]

The organizational structure of the State Committee of Standards is divided into administrations or departments, which carry out overall co-ordination of the drafting of standards for particular sectors of industry. The State Committee has also established a number of scientific-research institutes (probably about 20 of them) to assist its relevant administrations and departments by executing research work into problems of standardization and metrology, and checking the technical content of draft standards prior to approval. The practical work of drafting standards is delegated by the State Committee to research and development organizations already established within the appropriate industrial ministries and responsible to them for product and process development. These establishments are frequently referred to by the State Committee of Standards as 'base organizations for standardization', or 'head organizations for standardization', in those cases where a particular industry may be large enough to require close co-ordination of its base organizations. Over 400 research and technological organizations throughout the USSR have been designated in this manner.[35]

State standards are introduced into industrial practice by means of factory standards, published by the Standards Department of a particular enterprise and approved by its Chief Engineer. The technical requirements of these documents should coincide with the relevant state standards, but specify in more detail those quality parameters required for particular manufacturing and testing operations.[36] They are introduced into factory practice by the Standards Department which in a large enterprise may also supervise the work of small groups of standards engineers employed in different departments. In an attempt to ensure that all enterprises carry out this work, and hence produce articles to the quality level of the relevant state standard, a system of local inspection organizations ('laboratories of state supervision') responsible to the State Committee of Standards' Department of State Supervision has been established in all industrial areas throughout the USSR.[37] Their main duty is that of supervising the observance of relevant standards for finished products and measuring equipment by factories within their industrial location, but they also have the right to confiscate sub-standard production, through the authority of the office of the Chief State Inspector who is a Deputy Chairman of the State Committee of Standards.[38]

Most state standards specify major parameters of types of articles and relevant tests and acceptable tolerances to be observed during the product's manufacture and final test. These 'type standards', therefore, can be considered as useful instruments for establishing the major quality features of a product as manufactured. In practice, however, the quality in service, or the performance, of a product will also be dependent upon many other factors which it is difficult to standardize. In a Western-type economy, it is assumed that market pressures force manufacturers to take account of these factors during the design and manufacture of their product range in order to remain competitive; but in a tightly planned Soviet-type economy, with its associated seller's market, these market pressures are almost absent since purchasers have restricted choice in their buying activities. The USSR has consequently attempted to reduce this problem by means of a 'mark of quality' or 'quality attestation' system.

The main feature of the 'mark of quality' system is that it is a serious attempt to improve the quality of Soviet industrial production by granting an award to those products which are considered to meet the same requirements as similar advanced products sold by other manufacturers in the world market. The products for these awards are proposed by the factory and industrial ministry responsible for their manufacture, and tested for approval by an 'attestation commission' appointed by the manufacturing ministry and approved by the State Committee of Standards. This commission contains representatives of the manufacturers and users of the product, together with employees of the Ministry of Foreign Trade, if the product is considered to have some export potential, and a representative of the State Committee of Standards. If the product is successful, it is defined as being in the 'highest category of quality' and the producing factory receives a price increment.[39] If the product is not considered to meet this level, but still meets the relevant criteria of state standards, it is defined as 'first category of quality'. If the product fails this test, however, it is defined as 'second category', and under new regulations since 1985 must be removed from production within two months, unless special permission to continue production is granted by the State Committee of Planning for a two-year interval.[40]

The 'attestation' system consequently differs from 'type standardization' which attempts to stabilize the technical level of all factories producing a specific type of item. The 'mark of quality' system, on the other hand, attempts to create incentives for factories in a leading position in Soviet technology to manufacture products to the highest international levels. These levels may be higher than those specified by the product type standards, and include a detailed assessment of various product parameters, methods of manufacture and quality control which it may be too time-consuming to include in a 'type standard'. In addition, some assessment will be made of product style, and the degree of use of standard and common parts. The 'quality attestation' process, therefore, assists in attempting to guarantee a mix of good quality products being delivered to the national economy. In addition, under new procedures recently introduced into the USSR, quality attestation is to be carried out at the design, pre-production and final manufacturing stage of most industrial products for purposes of classification into the 'highest' or 'first' categories of product quality. Approval at the design and pre-production stages are hurdles which the product must clear before manufacture begins, but the final decision on quality attestation is still deferred until the product is manufactured under batch-production conditions.[41]

The first products to be awarded a 'mark of quality' were a series of electrical motors produced by the Moscow 'Vladimir Il'ich' Factory, and approved in April 1967. Since that time, Soviet publications have frequently cited evidence of the successful diffusion of the state attestation system throughout the Soviet national economy, particularly in engineering ministries and engineering factories.

It can be seen, therefore, that the Soviet authorities have made serious efforts to develop and implement a comprehensive system of standardization, and to use this system to stabilize and improve product quality in the economic conditions prevailing in the USSR. The following section of this chapter consists of a discussion of two particular case-studies of the use of state standards to evaluate Soviet product quality, to demonstrate the use of these openly-published documents to evaluate product parameters, and to contribute to the Western debate on Soviet product quality as outlined in the first section of this chapter. The particular cases considered are those of general-purpose machine tools and asynchronous electric motors. In the first case, type standards are used, whilst in the second case reference is also made to 'mark of quality' standards.

THE CASE-STUDIES – GENERAL PURPOSE MACHINE TOOLS, AND ASYNCHRONOUS SQUIRREL CAGE ELECTRIC MOTORS

General-purpose machine tools

General-purpose machine tools are widely used in almost every industrial sector in the USSR, as in the Western industrial economies, to produce engineering components in unit, small-batch and medium-batch production conditions. The basic requirement of any general-purpose machine tool, assuming that it has adequate overall capacity, is its capability to produce components of the required accuracy. This capability depends, in its turn, on the precision with which major elements can be moved and positioned in relation to one another. Thus, standards relating to machine-tool quality specify relevant tests and acceptable tolerances of error for the alignment of those major elements from which the machine tool is constructed.[42]

A study was initially carried out by Hill in 1969[43] to determine the range of products manufactured by the Soviet machine-tool industry, for which state standards had been published. Since it appeared from this survey that approximately 90–95 per cent of the Soviet output of general-purpose machine tools was specified by appropriate state standards, it was considered to be important to evaluate the technical requirements specified in these documents. Time did not permit a complete study of standards relating to every product type, and hence attention was concentrated on knee- and column-milling machines and centre lathes, as examples of widely used types of general-purpose metal-cutting machinery.[44]

The Soviet machine-tool standards were technically assessed by comparing the technical requirements embodied in these documents, with those adhered to by British manufacturers of similar machine models to which the Soviet standards related, since there were no relevant British national standards for milling machines and centre lathes at that time.[45] Consequently, the accuracy requirements specified in the Soviet documents (GOST 13–54 for milling machines and GOST 42–56 for centre lathes) relating to 1250 mm table-size milling machines and 400 mm swing-centre lathes, were compared to those adhered to by British manufactures of similar models. From this comparative assessment it was concluded that although there were many tests for which similar tolerances were specified by both the appropriate Soviet state standard and the British manufacturers' testing documents, there were also some important tests for which closer tolerances were required by the British-built machines.

A subsequent study[46] was made of later Soviet standards for the accuracy of milling machines and centre lathes published in 1972, for introduction into industrial practice at the beginning of 1974. These Soviet standards included tests and tolerances similar to those specified in the previous standards, although some important tolerances were also more demanding. The general trend was towards improving accuracy requirements, with many tests approaching the requirements of the British models selected for comparison. In addition to these accuracy requirements for 'machines of normal precision' (*stanki normal'noi tochnosti* ('N' class)), the new standards also contained provision for the requirements of machines conforming to 'improved precision' (*stanki povyshennoi tochnosti* ('P' class)) in which the alignment tolerances were usually 0.6 of those allowed for in 'normal precision' machines; whereas the requirements of 'improved precision' machines had previously been published in separate standards. Consequently, many of the test tolerances for these latter machines were more demanding than those of the Western models selected for comparison. Furthermore, in the case of centre lathes, tolerances were also included for 'high precision machines' (*stanki vysokoi tochnosti* ('V' class)) which were even more demanding than those for 'improved precision' machines.

The method used in the previously described studies consisted of a comparison of alignment tolerances specified in a Soviet state standard for a particular type and size of machine tool, with those adhered to by the manufacturer of a 'typical' British model of machine of the same type and size. In other words, the alignment tests for the British machine models were used as criteria against which to compare requirements of Soviet standards, in both the late 1960s and the mid-1970s.

In recent years, however, the British Standards Institution has published a series of standards relating to the accuracy requirements of many types of general-purpose machine tool. A comparison of the accuracy requirements specified in these standards with their selected Soviet counterparts should consequently form a better basis for assessment than the previously used 'selected typical model' approach, since national standards for similar machines can be compared, and both sets of standards use similar tests and metric tolerances. Furthermore, both sets of standards are up-to-date, and should consequently represent contemporary machine-tool standardization in both countries; and a sufficient time has elapsed to enable the requirements of the standards to be absorbed in both countries.

Consequently, it was initially decided to compare the alignment accuracies of knee- and column-milling machines and centre lathes, as specified in the relevant Soviet and British standards, in order to extend the previously described machine-tool studies. As these pilot studies showed that comparative analysis was possible using the Soviet and British documents, the methodology was extended for the following types of machines which had not been compared previously:

- radial drilling machines,
- vertical drilling machines,
- surface grinding machines,
- external cylindrical grinding machines,
- internal cylindrical grinding machines,
- vertical broaching machines,
- vertical gearhobbing machines,
- horizontal gearhobbing machines.

These machine types were selected since they process the bulk of general-purpose machining work carried out by the majority of industrial users. This machining work includes the generation of plain cylindrical and flat surfaces, which act as envelopes for the majority of engineering components, the generation of external helical gear teeth, and the generation of complex-shaped surfaces. In the light of available estimates it is considered that the selected types of machines account for 65 per cent of the total annual quantity of metalcutting machine tools produced in the Soviet Union, or more than 75 per cent of the total Soviet annual output of 'standard' (i.e. 'non-specialized') machine tools, for which it is feasible to publish standardized alignment accuracies. Numerically-controlled machines were not included in the study, since no Soviet standards had been published when the study was begun.

It is necessary to point out, however, that some further selection was carried out when comparing the alignment accuracies specified in each of these documents; this was because of the wide range of major product parameters used by the engineers responsible for drafting the appropriate standards. In spite of the necessity of this selection, we still consider that a valid comparison has been achieved over a wide range of diverse machine types.

Table 5.1 below is a summary of the results obtained from the comparative study referred to in the previous paragraph, and a full account of these results is available elsewhere.[47] Table 5.1 shows that for the total of 130 alignment tests relating to the ten machine-tool types and sizes selected for comparison (not counting surface grinding machines, since there were no directly comparable classes of machine accuracy), similar tolerances were specified for over 50 per cent of the tests in both the Soviet and British Standards, with the Soviet tests being more accurate in approximately 22 per cent of the cases, and the British tests being more accurate in 25 per cent of the cases. If surface grinding

Table 5.1 Comparison of accuracy requirements specified by British and Soviet standards for machine-tool alignment tests.

Machine type	Soviet standard more demanding (no. of occurrences)	Equal tolerances (no. of occurrences)	British standard more demanding (no. of occurrences)
Normal precision centre lathes	3	11	0
Improved precision centre lathes	4	10	0
Normal precision milling machines	0	16	0
Normal precision vertical drilling machines	3	1	0
Radial drilling machines	2	3	0
High precision surface grinding machines	8	0	0
Improved precision cylindrical grinding machines	3	7	3
Improved Precision Internal grinding machines	2	7	2
Vertical broaching machines	5	0	5
Normal precision vertical gearhobbing machines	1	6	7
Improved precision vertical gearhobbing machines	5	2	6
Normal precision horizontal gearhobbing machines	0	2	6
Improved precision horizontal gearhobbing machines	0	5	3
TOTALS (not counting surface grinding machines)	28	70	32
TOTALS (counting surface grinding machines)	36	70	32

machines are included, the proportion of tests for which the Soviet tests were more demanding is marginally higher (26 per cent), compared with 24 per cent for the British tests.

It can be concluded, therefore, that the alignment accuracies specified in Soviet state standards for centre lathes, and drilling, milling, grinding, broaching, and gearcutting machines are generally as high as those specified in British national standards. If it is assumed that these alignment accuracies are an indicator of quality for general-purpose machine tools, and that they are also an indicator of minimum quality requirements for the machine-tool industries in the respective two countries, then there is strong evidence to suggest that the quality of Soviet-produced general-purpose machines is similar to that of their British-manufactured counterparts. Furthermore, since many of the requirements of the British Standards are based on the recommendations of the International Standards Organization, and the UK machine-tool industry is a long-standing supplier to international markets, it would appear that the USSR can now be viewed as a potential international supplier of general-purpose machine tools of adequate quality.

It is necessary, however, to sound a note of caution over the use of these documents to obtain a total picture of product quality, since there may be several characteristics of a product which affect its quality in use, but which are difficult to specify quantitatively in a state standard. For example, if quality is viewed as total 'fitness for purpose', it is apparent that in addition to alignment accuracy, the 'fitness for purpose' of a general-purpose machine tool is also influenced by such factors as rigidity, ease and speed of setting, operation and control, and reliability and durability in service. These parameters can only be monitored by rigorous product-testing in laboratory and working conditions; but such a set of tests would be far more expensive to carry out than the procedures used in this current research programme. In this study, we have been concerned not only with establishing estimates of Soviet product quality, but also in achieving these estimates in the most inexpensive manner possible.

Moreover, some British machine-tool factories may also claim that market forces require their products to be far more accurate than the minimum requirements laid down by national standards, whereas such forces are considered to be absent from the centrally-planned Soviet economy. The validity of such a claim could only be verified by comparative analysis of individual machine models, however, which would require a very extensive study. In view of the comparatively recent publication of the British national standards, and the participation of the British machine tool industry in the drafting of these documents, we are inclined to the view that their requirements should be typical for the industry.

Asynchronous squirrel cage electric motors

More than three-quarters of all electrical motors are of the asynchronous type,[48] their widespread popularity arising chiefly from their design and physical construction. A simple and robust construction makes them suitable for high-speed operation and use in environments where severe operating conditions are encountered. The absence of slip-rings, brush gear and electrical connections to the rotor creates a motor of low cost, requiring the minimum amount of maintenance, and also means that during normal operation little or no sparking occurs, thus greatly reducing the risks of fire and explosion. Other inherent advantages include power factor ratings close to unity[49] and high values of operational efficiency, both of which depend on the rated power output and synchronous speed of the motor.[50] Furthermore, within defined operating ranges of torque and speed, the squirrel cage motor approximates to a constant speed machine. This feature, coupled with the motor's simple construction makes it ideally suited for building into machine tools in the form of a stator-rotor unit.[51] These motors do have certain disadvantages of large-starting currents and low-starting torques, which place limitations on their application, although design innovations have been introduced over the years to improve the motor's starting torque.[52] For the purposes of assessing the quality level of Soviet-manufactured asynchronous motors, the following parameters were selected as being representative of the parameters specified in Western and Soviet standards, and Western manufacturers' catalogues.[53]

> power factor;
> efficiency;
> percentage slip at rated load;
> ratio of initial starting current to rated current;
> ratio of initial starting torque to rated torque;
> ratio of maximum torque to rated torque;
> power to weight ratio;
> manufacturer's guarantee period;
> probability of fault-free operation;
> mean time between failures (MTBF);
> service life.

These parameters were recorded for a sample of 40 motors, 13 of which were manufactured in the USSR. The rated output for this sample ranged from less than 10 watts up to more than 1,000 kw, and this range was covered in 12 separate steps of power output groupings. At each step, it was ensured that motors selected for comparison were also of the same synchronous speed-rating, this having a direct bearing on the weights of individual machines, since machines with higher synchronous speeds have fewer pairs of poles and vice versa. Although information on all 15 operational parameters appears at least once in the tables, normally details on only seven to ten indices were obtainable, from either Soviet state standards or individual Western motor manufacturers. One of the prime reasons for this was a general reluctance, on the part of Western firms, to commit themselves fully on motor reliability information. In addition to the normal 12 months guarantee from the date of delivery, most were only prepared to provide the number of operational hours of the motors that consisted of (1) the bearing life expectancy; (2) the inter-servicing period of bearings; and/or (3) the period of time between replacement and lubrication of bearings. Explaining this, a Western company representative stated that as bearings were the only moving part of squirrel cage motors subjected to constant 'wear-and-tear', it was only necessary to issue reliability data on these components. The Soviet 'mark of quality' standards, on the other hand, were generally more consistent in their content on motor reliability.

A complete set of results for this comparative study is presented elsewhere,[54] whilst graphs 5.1 – 5.7 below show the values of seven selected major motor parameters for which comparative information was available across a whole selected range of motor output power. These seven parameters are:

> power to weight ratio (graph 5.1);
> efficiency (graph 5.2);
> power factor (graph 5.3);
> full load slip (graph 5.4);
> ratio of maximum torque to rated torque (graph 5.5);
> ratio of initial starting torque to rated torque (graph 5.6);
> ratio of starting current to rated current (graph 5.7).

In the case of the 'Western-built' motors, each group of motor power output represents the average value for all motors within that group (including three Polish-made motors distributed in the UK), whilst for 'Soviet-built' machines, the value is for one motor only.

Throughout the whole range of results no consistent pattern emerged to suggest that either Western or Soviet manufactured machines were technologically superior. It could be argued that, as far as 'power-to-weight' ratios are concerned, motors of Western origin hold a slight advantage in the upper and lower power groupings, whereas Soviet-built machines in the mid-range groupings appear to show better results; but even here care must be exercised before reaching any definite conclusions. For example, the results obtained for the 75 kw and 132 kw groupings indicate the Soviet AB2 and 3 series to be technically better than their Western counterparts, although further reading into the contents of the standard[55] concerned suggests otherwise. First, no obligations are placed with the manufacturer to equip the motors with damp-proof, oil-resistant or chemically-protective insulation, and second, the machines are not expected to be of explosion-proof construction. These two factors obviously limit the practical applications of the machines to non-hazardous working environments and temperate climates, and this is confirmed by the relevant specifications[56] stipulated for these motors in GOST 15150–69.[57] Any modifications or design changes aimed at rectifying these limitations are certain to have an adverse effect on any 'power-to-weight' advantage the motors possess at present. Most of what has just been said applies also to the Series A2 motor range of the Moscow 'Vladimir Il'ich' factory.

With respect to the other major parameters, some differences do exist but, generally, the large variations in values that occur tend to be in a small number of cases. The close similarity in the results appears to be a recurrent theme throughout the whole range of power groupings and is even more in evidence when comparing the technical requirements for rotating electrical machinery in general. The variations from the rated norms and the tolerance limits, within which both Soviet- and Western-made motors are permitted to operate, are identical for 25 technical requirements listed in BS4999 parts 30, 32, 60 and 69 and GOST 183–74.[58] As far as the more specialist motor[59] was concerned,

Product Quality, Standards and Technical Progress

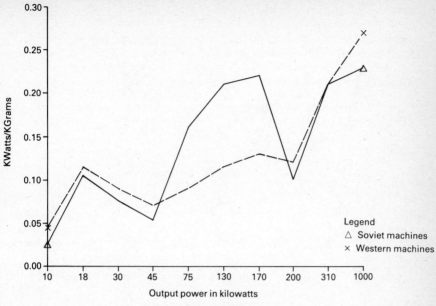

GRAPH 5.1 *Power to weight ratio*

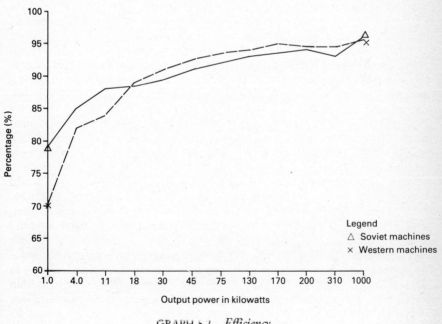

GRAPH 5.2 *Efficiency*

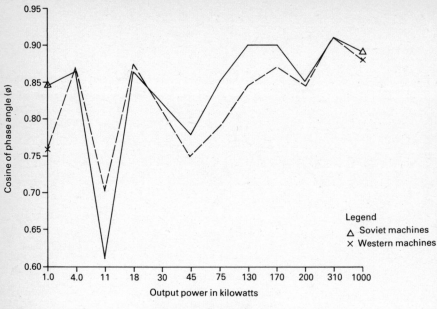

GRAPH 5.3 *Power factor*

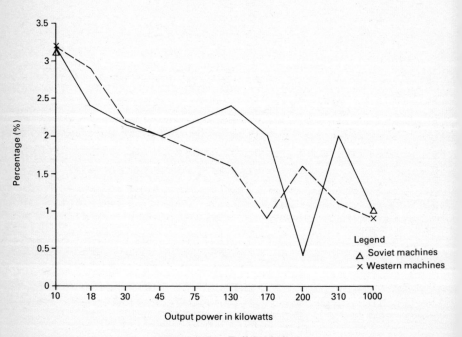

GRAPH 5.4 *Full load slip*

Product Quality, Standards and Technical Progress

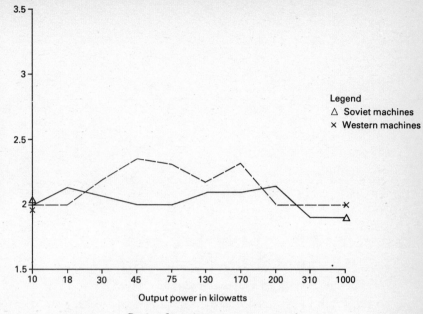

GRAPH 5.5 *Ratio of maximum torque to rated torque*

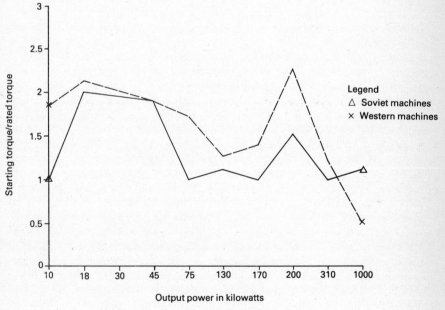

GRAPH 5.6 *Ratio of starting torque to rated torque*

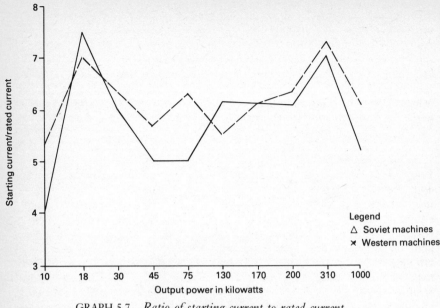

GRAPH 5.7 *Ratio of starting current to rated current*

similarities existed for the values in the relationships of Starting Torque to Rated Torque and of Starting Current to Rated Torque, as specified in tables 1 and 2 of GOST 10799–77, and part 11 of BS5000, which caters for small-powered electric motors and generators.

From a theoretical point of view at least, it appears that very similar demands and stringencies are placed on both Soviet and Western motor manufacturers in their design, construction and testing of squirrel cage machines, and that to make a proper assessment of the capabilities of individual motors, closer scrutiny of the relevant standard is necessary (similar to that which was carried out on the series AB2 and 3 motors above). In practice though, perhaps the best yardstick by which to judge the qualities of a motor is its ability to maintain consistently its rated values of performance. In this respect, much depends on the working environment within which the motor is placed and whether it is properly equipped to cope with the prevailing conditions of operation. In the Soviet Union, incorrect application of asynchronous motors accounts for between 15 per cent and 35 per cent of their failures,[60] which, according to the same source, was approximately two to three times the number of failures which resulted from the natural causes of ageing and fair 'wear-and-tear'. Deficiencies in design and production were another major source of failures (30–35 per cent) of Soviet asynchronous motors, as were operational deficiencies and poor quality of repair (35–50 per cent). Under the heading 'operational deficiencies' we should also include those points highlighted in a recent paper by Julian Cooper,[61] namely the inconsistency of voltage regulation and inferior quality of power supplies.

As a final point, it is interesting to note that Ermolin and Zherikhin[62] mention that the degree of airgap evenness between rotor and stator represented an important reliability indicator for asynchronous motors. It was also admitted that the relevant state standard makes no provision for the checking of this gap due to the practical difficulties involved. Indirectly, it seems that these statements justify the importance given by Western manufacturers to the reliability of bearings; they also hint at the possibility that responsibility for overall motor reliability and quality of production lies within the realms of other branches of engineering which are beyond the terms of reference of this study.

COMMENTS AND CONCLUSIONS

From the results of the research programme described in this chapter, it is apparent that the tests and tolerances specified in Soviet state standards for machine tools and asynchronous squirrel cage electric motors are generally equivalent to those specified in their British counterparts. In view of these results, therefore, it would seem to be appropriate to reconsider some of the generally held Western views of Soviet product quality, since some of the criticism may be overstated. It could be argued, however, that the selection of machine tools and squirrel cage motors is a very small, and perhaps biased sample, in view of the fact that machine-tool manufacture and electrical engineering have been priority civilian industries in the USSR since the early 1930s, and to base conclusions on such a sample is not totally justifiable. To counteract such a criticism, therefore, the research methodology is currently being extended to cover the comparative study of other product groups in the industrial and consumer sectors, including refrigerators, photographic equipment and motor cars, and these results will be published in due course.

Further research should also establish the extent to which the development and implementation of a standardization policy in the USSR has served to counteract some of the factors which are claimed to have led to poor product quality during the rapid Soviet industrialization drive. It is apparent that for many years, the leading Soviet industrial policy-makers have attempted to use standardization procedures in the context of an industrializing and expanding planned economy, to contribute towards the stabilization and improvement of Soviet product quality; and that these procedures have been made more rigorous since 1965. Furthermore, it is apparent that the methodology of standardization has lent itself very well to 'plan fulfilment' conditions prevailing in a planned economy, since standardization quantifies a product's major parameters and tolerances as a basis for assessment. As the national economy became more sophisticated, however, type-standardization was supplemented in the late 1960s with quality attestation which aimed at quality improvement of individual products made at individual factories, assessed on a broader basis than 'major parameter fulfilment'. According to the results presented in this chapter, these policies and procedures appear to have been successful in the machine-tool and electrical motor sectors at least, and further research should test whether the same is true for a broader range of product groups. It is also interesting to speculate whether type-standardization and quality attestation may also be supplemented by 'certification', a policy followed in some Western countries where individual factories are certified as capable of achieving specific quality levels for defined product types.[63] This would represent a move from 'product-based' approval to 'enterprise-based' approval, across a broader range of products than those considered in individual 'mark of quality' standards; it may be an appropriate procedure for the Soviet economy as its growth towards sophistication continues.

It is clear, however, that even if the results of further research are similar to those obtained from the study described in this paper, the criticisms of soviet product quality made by Western scholars and outlined in the first section of this paper should not be conveniently dismissed. In the first place, the views of most Western scholars have been formed from close study of a broad range of Soviet political and economic publications, in which the discussion on product quality is usually frank and open. The context of many discussions in the Soviet press needs to be kept firmly in mind, however: namely, the creation of an atmosphere of debate which can be frank and critical of current shortcomings in order that steps be taken towards improvement. Furthermore, some criticisms relate mainly to the inconvenience of presentation of Soviet standards, and the apparently trivial nature of some changes in standards requirements.[64] It is not unusual, though, to hear Western designers express similar sentiments about standards used in their own organizations. It is suggested, therefore, that the research methodology based on comparative technical assessment as described in this chapter, forms a useful contribution to an evaluation of Soviet product quality alongside the published sources used by other Western scholars. Such quantitative material on Soviet product quality can be seen alongside other data on the economic, political and administrative factors influencing product quality and development, as presented by David Dyker in chapter 8.

The methodology described in this chapter may also be criticized for concentrating almost exclusively on a product's major parameters, whereas the total performance of a product in service may also be influenced by other, more minor, parameters. Many of these latter parameters are frequently difficult to standardize, and some are often discretionary; but the aggregate effect of these minor parameters can have a significant effect on product performance. The 'quality attestation' system is an attempt to ensure satisfactory conformance of all of a product's parameters, however, and clearly this system is worthy of further research, particularly as steps have recently been taken by the Soviet authorities to improve the system.[65]

As a final point for discussion, it is important to note that the content of the documents discussed in this present chapter has referred to products at the mature stage of the product life-cycle, in view of the documents available. Several research reports have been written by Western scholars, however, which point to certain shortcomings in the stages of product and process innovation in the USSR, giving rise to long time-cycles in product and process development.[66] Thus it would be useful to study the degree to which standardization may influence the processes of technical development in the Soviet economy, since more standards for more advanced technology products may become necessary for an increased rate of innovation in the USSR.

6
Technology Transfers and Technology Controls: a Synthesis of the Western–Soviet Relationship*

GARY K. BERTSCH

The transfer of technology from Western industrialized countries to the Soviet Union, and Western efforts to control this transfer, have been highly visible and contested issues in the 1980s. Some observers – notably those within certain quarters of the Reagan Administration – have called attention to a Soviet 'technological raid', a 'haemorrhage of a national heritage', and to significant Eastward transfers that help prop up a decaying Soviet economy and contribute to the Soviet military build-up. Others take more moderate positions; although recognizing that some technology is being transferred from the industrialized Western countries to the Soviet Union, they note that the amount of the transfers are not particularly large, given the size of the Soviet economy; that the impact of the technology is limited, given Soviet difficulties in assimilating and diffusing advanced technology; and, that the Soviet military build-up has been conducted largely independently of Western technology. These contrasting viewpoints, in turn, suggest rather different Western responses to the issue of technology controls. The former tend to feel that more restrictive controls are needed to stem the technological haemorrhage; the latter take a more relaxed view arguing that although some Western controls on military relevant transfers are advised, the most significant barriers to technology transfer are those systemic obstacles that the Soviets impose upon themselves. To shed more light on these and related issues, this chapter attempts to synthesize current opinion and research concerning the nature and impact of Western–Soviet technology transfer, and, to examine recent Western efforts and controversies surrounding the control of these transfers. I shall argue that although considerable attention has been devoted to the issues of West–East trade and technology transfer over the last decade, most of the analyses of the really significant issues – for example, the impact and consequences of technology transfers – remain tentative and ambiguous. This ambiguity and uncertainty in scientific research and public opinion leaves policy-makers considerable freedom of action when they address the issue of technology controls. Accordingly, Western policies on technology and trade controls are more often based upon national political and economic considerations than upon scientific assessments of the importance of technology transfer in the East–West economic, technological and military balance.

*The author would like to acknowledge the support of a NATO Research Fellowship; the Fulbright Program; the University of Georgia, USA; and the University of Lancaster, UK, in facilitating the research on which this chapter is based.

TECHNOLOGY TRANSFER: DEFINITIONS, FORMS, AND IMPACT

Discussions of technology transfer always raise a host of questions about definitions. Are we talking about the transfer of products, know-how, increased productivity, or some other phenomenon? Discussions of Western–Soviet technology transfer generally involve all of these phenomena. Philip Hanson refers to international technology transfer as a 'process whereby the productivity of resources of one country can be increased by the transmission from other countries of information or of products and processes embodying that information'.[1] Hanson uses the term in the comprehensive sense to include transfers deliberately arranged by the supplier of the technology and those not intended by the supplier, and those that raise the productivity of resources and those that do not. In so doing, he suggests that transfers be identified by their potential to raise the productivity of resources in the recipient nation, not by their actually doing so.

In this chapter, technology transfer will be used in its most comprehensive sense. It will refer to the transfer of products, information, processes, and know-how. Sometimes these transfers will result in increases in the productivity of the Soviet Union; in many cases they will not. In this regard we should call attention to the considerable literature noting Soviet difficulties in utilizing, assimilating and diffusing advanced technology.[2] Because of the systemic difficulties, transfers of products, information, processes, and even know-how do not always achieve the expected benefits from the Soviet perspective, or the perceived costs, say from the Western national security perspective. That is, transfer of the potential may have less significant consequences than those expected by both the recipient and supplier. As a result, it should be noted that our and other comprehensive uses of the term 'technology transfer' may cause one to overestimate the real consequences of the transfer process. Yet, because it is often difficult to assess just how successful (i.e., in terms of increased productivity) transfers really are, we will risk the danger of overstating the real significance of the transfers in order to make the analysis as comprehensive as possible. When it is possible to say that the transfer of certain products or know-how, are known on the basis of accepted research to have had a discernible impact, we will do so. More often, unfortunately, we will have to conclude that although we are aware that some potential transfer took place, we find it difficult to assess precisely the real impact. Hence we use the comprehensive conception of technology outlined above.

In this age of rapid technological change, expanding communication and an open world economy, West–East technology transfers take a multiplicity of forms. Some of these forms are relatively public and amenable to examination and measurement; others are covert, and by definition, much more difficult to assess. Although we are unable in a chapter of this length to review and assess all of the research and opinion on West–East transfers, it is necessary to try to highlight what we do and do not know, particularly as it relates to the question of technology controls which will be addressed in the second part of the chapter.

Beginning with those forms that are more open – that is, what will be called the commercial, official, and legal transfers – we can observe that the opportunities for West–East technology transfer have increased markedly over the last two decades.[3] In the 1960s Soviet and East European decision-makers decided to take greater advantage of the commercial and technological opportunities provided in the West. Western exporters, scientists, and governments (with some important exceptions, noted below) were eager to expand commercial, scientific and technological relations with their Eastern counterparts. Among the consequences were major increases in Western sales and Soviet (and Eastern European) purchases of capital goods and know-how.

The US Department of Commerce (DOC) has attempted to quantify the sale of Western high technology to the Soviet Union and other Communist countries.[4] Using the standard international trade classification (SITC) to define some 30 categories of high technology, including machine tools, computers, jet and gas turbines and so forth – categories the researchers consider likely to contain products embodying world best practice in critical technologies – the DOC study results in a number of interesting observations. First, Communist countries altogether import a relatively small share of the industrialized West's[5] high technology exports, and the share has been falling (from 4.7 per cent in 1970 to 4.1 per cent in 1981). Second, the share of Western high technology exported to the Soviet Union is relatively small and is, in fact, somewhat below the share of Western high-technology exports going to the rest of the world (see table 6.1). The data for 1981 indicate that Western high technology exports to the Soviet Union accounted for 12 per cent of Western manufactured goods exported to the Soviet Union, and 8.3 per cent of total exports; the respective shares of Western high-technology exports to the world were somewhat higher, 16.4 per cent and 12.4 per cent. The respective shares to Eastern Europe were 15.6 per cent and 10.7 per cent; to all Communist countries they were 14.0 per cent and 9.8 per cent. Therefore, one might conclude that the amount of technology transferred to the Soviet Union and other Communist countries through commercial sales has not been particularly great, compared with the shares of Western technology exported to the rest of the world.

The DOC study also provides data on the Western sources of Soviet high-technology imports. In 1981 the Federal Republic of Germany was the largest Western exporter to the Soviet Union and accounted for 28.9 per cent of Western high technology exports (see table 6.2). The second largest exporter, Japan, increased its high technology sales to the Soviet Union significantly over the 1970s, from about 11 per cent of the West's total in 1970, to over 21 per cent in 1981. The US and UK shares both declined, on the other hand, from 13.3 per cent and 13.9 per cent in 1970, to 3.3 per cent and 5.4 per cent in 1981, respectively.

Finally, the DOC study tells us some interesting things about the composition of these Western–Soviet technology transfers. The top five items in 1980 and 1981 for both the Soviet Union and Communist countries altogether were:

(1) machine tools for metal work;
(2) pumps and centrifuges;

Table 6.1 Comparison of high-technology exports with manufactured goods and total goods exports – 17 Western industrial* countries to the Communist countries and to the world: 1970, 1979, 1980, 1981 (US $ million)

Western industrial exports to:	1970 $	1970 % of	1979 $	1979 % of	1980 $	1980 % of	1981** $	1981** % of
USSR								
High-technology	402.9		2371.3		2330.3		1735.5	
Manufactured goods	2212.4	18.2	13642.4	17.4	15113.1	15.4	14435.4	12.0
Total	2490.8	16.2	18114.3	13.1	19837.5	11.7	20854.3	8.3
Eastern Europe								
High-Technology	414.0		2360.1		2194.2		1721.3	
Manufactured goods	2758.7	15.0	13802.8	17.1	14138.5	12.7	11040.0	15.6
Total	3522.7	11.8	18154.3	13.0	19460.9	11.3	16135.6	10.7
Total all Communist countries								
High-technology	1172.4		6936.7		6958.0		5649.3	
Manufactured goods	8009.5	14.6	43360.1	15.9	46443.6	15.0	40398.9	14.0
Total	9521.9	12.3	55392.5	12.5	61928.9	11.2	57468.7	9.8
World								
High-technology	24770.9		118135.6		136205.3		144480.8	
Manufactured goods	162940.1	15.2	768044.2	15.4	892324.8	15.3	881911.7	16.4
Total	211644.5	11.7	999173.8	11.8	1173144.0	11.6	1161478.0	12.4

* Western industrial represents 17 industrial Western states (USA, Canada, Japan, Belgium–Luxembourg, France, Federal Republic of Germany, Italy, Netherlands, Austria, Norway, Sweden, Switzerland, UK, Denmark, Finland and Ireland.
** 1981 data include only estimates of UK exports.
Source: Department of Commerce from UN Series D Trade Data.

Table 6.2 USSR sources of Western industrial high-technology products (US$ million)

	1970		1972		1979		1980		1981	
	High-technology exports to USSR	As % of total	High-technology exports to USSR	As % of total	High-technology exports to USSR	As % of total	High-technology exports to USSR	As % of total	High-technology exports to USSR	As % of total
Canada	0.2	—	0.7	0.2	11.3	0.5	27.5	1.2	0.4	0.0
USA	12.5	3.1	26.8	8.1	154.7	6.5	84.7	3.6	56.5	3.3
Japan	43.5	10.8	41.8	12.7	398.9	16.8	400.2	17.2	366.0	21.1
Belgium–Luxembourg	5.9	1.5	4.3	1.3	21.1	0.9	18.0	0.8	12.1	0.7
Denmark	4.8	1.2	4.6	1.4	17.1	0.7	23.1	0.1	17.9	1.0
France	58.5	14.5	38.0	11.5	376.8	15.9	341.3	14.8	204.7	11.8
Federal Republic of Germany	92.9	23.0	79.4	24.1	668.3	28.2	727.2	31.6	501.8	28.9
Ireland	—	—	—	—	1.3	0.1	0.2	Negligible	0.0	0.0
Italy	69.6	17.3	51.1	15.5	257.2	10.8	222.2	9.6	156.3	9.0
Netherlands	1.1	0.3	3.2	1.0	16.7	0.7	6.1	0.3	10.0	0.6
UK	56.0	13.9	43.4	13.2	94.0	4.0	125.7	5.5	93.6*	5.4
Austria	5.6	1.4	4.8	1.5	43.7	1.8	48.2	2.1	30.4	1.8
Finland	6.3	1.6	5.4	1.7	70.9	3.0	86.2	3.7	121.8	7.0
Norway	0.1	—	0.2	0.1	8.2	0.3	12.3	0.5	6.5	0.4
Sweden	22.3	5.5	9.0	2.7	119.9	5.1	71.1	3.1	77.3	4.5
Switzerland	23.6	5.9	16.3	5.0	111.3	4.7	136.4	5.9	80.0	4.6
Total	402.9	100.0	329.2	100.0	2371.3	100.0	2330.0	100.0	1735.5	100.0

* Estimated
Source: US Department of Commerce from UN Series D Trade Data.

(3) electrical machinery;
(4) electrical measuring and control instruments;
(5) cocks, valves and related items.

These five items represented 63.9 per cent of the 1980 total for Western technology exports to all Communist countries. The surprising and significantly lower exports of electronic, communications, and aircraft categories no doubt reflects the restrictive influence of Western strategic trade controls.

Another important official form of Western–Soviet technology transfer has involved inter-governmental agreements on scientific and technical co-operation. In the early 1970s, for example, the USA and the USSR entered into 11 separate agreements pledging co-operation in the fields of:

(1) science and technology (1972);
(2) environmental protection (1972);
(3) medical science and public health (1972);
(4) space (1972);
(5) agriculture (1973);
(6) world oceans (1973);
(7) transportation (1973);
(8) atomic energy (1973);
(9) artificial heart research and development (1974);
(10) energy (1974);
(11) housing and other construction (1974).

It should be noted that the USA reduced funding and other support for these exchanges, following the Soviet invasion of Afghanistan, and allowed activities to decline further in 1982 in response to Soviet complicity in the imposition of martial law in Poland; however, American support for the exchanges and co-operation in the programme picked up in 1983, and the exchanges assumed more normal patterns in the mid-1980s. Although it is beyond the scope of this chapter to assess the technology transferred through these exchanges, others have conducted a variety of reviews.[6] In the late 1960s and 1970s, many other Western governments also entered into similar agreements pledging scientific and technical co-operation with the Soviet Union, including:

(1) France (1966, 1973);
(2) United Kingdom (1974);
(3) Italy (1974);
(4) Sweden (1970);
(5) Germany (1973, 1978);
(6) Finland (1974, 1975, 1977);
(7) Canada (1971);
(8) Japan (1973);
(9) Australia (1975).

Although there has been little systematic research assessing the consequences of these agreements, all observers acknowledge that they have expanded the opportunities for West–East technology transfer.

Another form of Western–Soviet technology transfer that is open to some examination, but still extremely difficult to measure and assess, involves the extensive and complicated modes of industrial co-operation between Western firms and their Soviet counterparts.[7] Industrial co-operation has taken a variety of forms and includes:

(1) sales of equipment (sometimes for complete production systems or turnkey plants) including technical assistance;
(2) licensing of patents, copyrights, and production know-how;
(3) franchising of trademarks and production know-how;
(4) purchasing and selling between partners involving exchanges of industrial raw materials and intermediate products;
(5) subcontracting involving the provision of production services;
(6) sale of plant, equipment, and/or technology with payment in resulting or related product;
(7) production contracting involving agreement for transferred production capabilities in the form of capital equipment and/or technology;
(8) co-production agreements allowing partners to produce and market the same products resulting from a shared technology;
(9) joint research and development, and a variety of other forms.[8]

These and other types of industrial co-operation are complex, often quite secretive, and usually difficult to assess. On the basis of a number of governmental and academic surveys, however, one must conclude that these forms of technology transfer have provided significantly increased opportunities for the transfer of information, processes and know-how.[9] Significantly, however, uncertainty remains, on the consequences of these transfer opportunities and potential.

Other forms of technology transfer that are even more difficult to assess involve foreign travel by scientists, participation in academic and scholarly conferences, and literature screening by Soviet specialists. Some in the US Government consider these forms of technology transfer as having particularly significant implications for national security. The CIA has claimed that more than one-third of the young Soviet scholars proposed to visit the USA under official exchanges in the early 1980s were 'completely unacceptable in terms of prospective technology loss'.[10] In their opinion, many of the exchange programmes needed to be modified or have Soviet access constrained before the exchanges could be allowed to continue. Accordingly, in the early 1980s, the USA Government showed increasing concern about the national security implications of these transfer channels, and paid closer attention to the research interests of Soviet and Eastern European visitors, and in some cases, restricted their participation in exchanges that were formerly more open to foreign participation. Some restrictions were opposed by influential members of the US academic community (I shall discuss this later in the chapter).

Literature-screening is a significant form of technology transfer that the Soviets have pursued for many decades. Given the vast quantities of scientific and technical publications in the West, and Soviet capabilities in reviewing them, one suspects that this is a primary source of Western technology. Although considered a 'passive' and therefore less effective mode of transfer –

as compared to the more 'active' and effective modes which involve considerable personal interchange – literature-screening provides significant opportunities for Soviet acquisition efforts.

Even more troublesome forms of Western–Soviet technology transfer involve the illegal and covert activities that often are thought to be of considerable national security consequence to the West and that are, at the same time, more difficult to discern and assess. Many serious allegations were made in the early 1980s. Admiral B. Inman, Deputy Director of the CIA until 1982, told a US Congressional Committee that 'as we look at the military useful, military related technology that the Soviets have acquired from the West, about 70 per cent of it has been accomplished by the Soviets and the East European intelligence services. They have used clandestine, technical, and overt collection techniques. . . .'[11] Among these forms of covert transfer are those carried on by Soviet and Eastern European diplomats. Interestingly, the number of Soviet diplomats expelled from Western countries, mainly for industrial and technical espionage, increased markedly in the early 1980s from 27 in 1981, to 49 in 1982, to roughly 100 in 1983. This was evidence that espionage was increasing, or that Western governments were more resolute in opposing such activities, or both.

Another form of illegal transfers involves diversion of controlled technology sales from legitimate trade destinations to proscribed destinations in the Soviet Union or Eastern Europe. There have been some high-profile, sensationalized cases in the 1980s, involving numerous middlemen engaged in various forms of profitable impropriety.[12] In view of the prices the Soviet recipients are apparently willing to pay for this technology (four to seven times the normal Western market price as reported by one US official),[13] such technology transfers are apparently highly prized in the Soviet Union. Although this chapter will not catalogue the extensive cases of known diversion and illegal sales (or speculate about the size and significance of the unknown cases), one should acknowledge this as a significant form of West–East technology transfer in the open world economy, and one requiring more careful attention than it has as yet received.

Another troublesome form of West–East technology transfer involves Soviet- and Eastern European-owned firms in the USA and the West. Noting that there were 20 such firms in the USA, and more than 300 in Western Europe by the late 1970s, the CIA alleges that these firms purchase controlled technology and study it in various Western countries.[14] Technically they are not in violation of Western export control laws unless they attempt to export the equipment or related technical data to the East without a licence. The CIA also claims that the Soviets get access to controlled microelectronic technology by setting up dummy corporations in the West which purchase sophisticated microelectronics equipment, which is then illegally shipped and reshipped to the Soviet Union or Eastern Europe.[15] Because of the relevance of these and other claims about technology flows to the issue of technology controls (to be discussed in the second part of this chapter), it is necessary to examine some of the leading opinion and research concerning the question of impact.

An impressive body of research has attempted to assess the impact of Western technology transfers on the Soviet and Eastern European economies;

unfortunately, many of the findings tend to be tentative, ambiguous, and sometimes difficult to interpret. First, there is a body of econometric studies that report somewhat different estimates of the contribution of Western technological exports to Soviet industrial growth.[16] The estimates range from those that say that Western technology contributes no more than indigenous technology, to those that say that Western technology contributes as much as 1 per cent per annum more. The range for the estimates is quite wide, and therefore difficult, particularly for the broader public and policy-makers, to interpret. However, among the important contributions of the econometric studies is a better realization of the difficulties in assessing the economic impact of Western technology, and a fuller appreciation of how little we really know. This is not to belittle the significant efforts of those involved in such assessments, but to draw attention to the difficulties encountered and the present state of the art.

A second body of literature utilizes case-studies to look at technology transfers and their impacts by particular countries, sectors, and projects. A noteworthy example is Philip Hanson's examination of the impact of Western technology on the Soviet mineral fertilizer industry.[17] Hanson describes and analyses the extensive Soviet use of imported Western plant and know-how in the production of nitrogenous and compound fertilizer. He acknowledges that the impact of the imports from the West on the production of mineral fertilizer supplies and Soviet agricultural output is difficult to assess because of the fact that such supplies and output are influenced by a variety of complicated factors. For example, contributory factors to increased Soviet crop output might be machinery, land and its care, improved seeds, pesticides, and so forth. However, Hanson concludes that there are still some important things that can be said on the basis of his research:

Incomplete and heavily guesstimated though it is, the evidence favors the view that imports of Western mineral fertilizer plants have been well-chosen and highly cost-effective. Given the salience of a backward agriculture and a backward chemical industry, the planners may not have too much difficulty in homing in on an area of technology importation that yielded high returns. However, the blending of indigenous, East European, and Western inputs in the various segments of the industry, and the rapid growth of capacity and nutrient supply, suggest that detailed implementation of this policy has been quite skilfully managed.[18]

A careful review of the many other case-studies reveals that few provide detailed and systematic analyses of the questions of Soviet and Eastern European assimilation and impact. One recent review concludes that the existing studies are 'tentative' about these issues and suggests that 'generalizations about these questions should be made with great caution'.[19] The author notes:

One could use the findings of the case studies to support the proposition that Eastern projects most often assimilate Western technology slowly and inefficiently in comparison with projects in the industrial West. In some cases, however, the case studies have provided evidence of relatively rapid and efficient assimilation.[20]

Despite their ambiguous and tentative nature, however, the author concludes that it may still be useful to summarize the case-studies' findings:[21]

(1) During the last two decades, the Soviet Union and East European countries

took considerable advantage of long-term 'active' technology transfer mechanisms in West–East trade.
(2) Sectoral studies reveal that Western technology has made significant contributions to the technological advancement of the Eastern industrial branches studied.
(3) However, country studies find little evidence that Western technology has had a major beneficial impact on the economies as a whole.[22]
(4) Furthermore, when compared with the experiences of Western industrialized countries, assimilation of Western technologies has been slow and inefficient in the East, apparently because of infrastructural and systemic shortcomings. After these and related observations, the author cautions that 'It should be re-emphasized that there are significant differences among the authors of the existing studies.'[23]

Again, the differences, ambiguity and tentativeness of the findings are not cited to belittle the case-studies. For the most part, these are impressive studies conducted by the West's most competent scholars. The findings are cited for another reason. This is to emphasize how little we still know about the important issues of technology transfer and to demonstrate the paucity of accepted, scientifically-based conclusions upon which Western trade-control policy might be formulated. Our understanding of the military impact of West–East technology transfer is even more tenuous.

Accordingly, what are the impacts of Western technology transfers on Soviet military capabilities? Again, this is a very large and complex question about which much is being said and written. No effort will be made here to review and assess all of the relevant research and opinion; rather, an attempt is made to summarize and highlight the issues and debate as they relate to the matter of technology controls.

First, there have been prominent individuals – particularly evident in some quarters of the Reagan Administration – who contend that West–East technology transfer is making major contributions to Soviet military capabilities, largely as a result of the unwitting liberality of the Western countries. US Secretary of Defense, Caspar Weinberger, for example, has warned that the Soviets 'have organized a massive, systematic effort to get advanced technology from the West. The purpose is to support the Soviet military buildup.'[24] A Department of Defense (DOD) white paper on *Soviet Military Capabilities* called attention to the Soviet's quest for technological superiority and the shrinking Western lead in applied military technologies.[25] The CIA study and the sanitized publication *Soviet Acquisition of Western Technology* argued that the Soviet Union and their Warsaw Pact allies have obtained vast amounts of militarily significant Western technology and equipment through legal and illegal means (see table 6.3). The published CIA report claimed that 'the Soviets and their Warsaw Pact allies have derived significant military gains from their acquisitions of Western technology, particularly in the strategic, aircraft, naval tactical, microelectronics, and computer areas.'[26] The report argued that Soviet acquisitions of Western technology have allowed the Soviets to:

(1) save hundreds of millions of dollars in R&D costs, and years in R&D

Table 6.3 Selected Soviet and East European legal and illegal acquisitions from the West affecting key areas of Soviet military technology.

Key technology area	Notable success
Computers	Purchases and acquisitions of complete systems designs, concepts, hardware and software, including a wide variety of Western general purpose computers and minicomputers, for military applications.
Microelectronics	Complete industrial processes and semiconductor manufacturing equipment capable of meeting all Soviet military requirements, if acquisitions were combined.
Signal processing	Acquisitions of processing equipment and know-how.
Manufacturing	Acqusitions of automated and precision manufacturing equipment for electronics, materials, and optical and future laser weapons technology; acquisition of information on manufacturing technology related to weapons, ammunition, and aircraft parts including turbine blades, computers, and electronic components; acquisitions of machine tools for cutting large gears for ship propulsion systems.
Communications	Acquisitions of low-power, low-noise, high-sensitivity receivers.
Lasers	Acquisitions of optical, pulsed power source, and other laser-related components, including special optical mirrors and mirror technology suitable for future laser weapons.
Guidance and navigation	Acquisitions of marine and other navigation receivers, advanced inertial-guidance components, including miniature and laser gyros; acquisitions of missile guidance subsystems; acquisitions of precision machinery for ball bearing production for missile and other applications; acquisitions of missile test range instrumentation systems and documentation and precision cinetheodolites for collecting data critical to postflight ballistic missile analysis.
Structural materials	Purchases and acquisitions of Western titanium alloys, welding equipment, and furnaces for producing titanium plate of large size applicable to submarine construction.
Propulsion	Missile technology; some ground propulsion technology (diesels, turbines, and rotaries); purchases and acquisitions of advanced jet engine fabrication technology and jet engine design information.
Acoustical sensors	Acquisitions of underwater navigation and direction-finding equipment.
Electro-optical sensors	Acquisition of information on satellite technology, laser rangefinders, and underwater low-light-level television cameras and systems for remote operation.
Radars	Acquisitions and exploitations of air defense radars and antenna designs for missile systems.

Source: CIA, Soviet Acquisition of Western Technology, 1982, p. 7.

development lead time;
(2) modernize critical sectors of their military industry and reduce engineering risks by following or copying proven Western designs, thereby limiting the rise in their military production costs;
(3) achieve greater weapons performance than if they had to rely solely on their own technology;
(4) incorporate countermeasures to Western weapons early in the development of their own weapon programs.[27]

Concerns about the military implications of West–East technology transfer have also come from other nations and quarters in the West. The French intelligence services have documents which purportedly show that the Soviet Union has made major strides in military modernization as a result of its acquisitions of Western technology.[28] The documents include material prepared by the KGB and the Soviet military industries commission (the VPK). One involves a detailed Soviet assessment emphasizing the value of Western technology to the Soviet Aviation Ministry. The document says that the Ministry received 156 samples and 3,896 specialized technical documents from Soviet intelligence agents stationed in the West. Of these, 87 samples and 346 technical documents were purportedly put to use in research and development of new military equipment and weapon systems. These acquisitions were reported to represent savings of 48.6 million rubles in 1979 and to have resulted in technological improvements to the SU-27 and Mig 29 Soviet aircraft.

Others in the West take a somewhat less sinister position on the question of the contribution of Western trade and technology to Soviet military capabilities. In a careful examination of Western technology and the Soviet defense industry, Julian Cooper argues that although it is indisputable that the Soviet military has benefited from Western technologies, 'the scale of such military-related transfers and their significance to the strengthening of the Soviet military potential have been widely overstated in the recent discussion, with an associated underestimation of the Soviet Union's capability for independent technological development.'[29] Although acknowledging that his conclusions are provisional, Cooper establishes that the contribution of Western technology to Soviet defense industries and military capabilities is much less than some would have us believe. Thane Gustafson also takes a sceptical look at the 'direct, near-term' military implications of West–East trade and technology transfer, arguing that 'the chances of a sudden doomsday giveaway through trade are nil.'[30] Although feeling that assertions about 'indirect, long-term' technological consequences are more credible, Gustafson maintains that Western technology will not be of more than passing benefit to the Soviet Union unless they develop their own capacities for technological innovation, generation, and diffusion.

Research and opinion on these and related questions concerning the consequences of West–East technology transfer remain unsettled. Without doubt, the divided, ambiguous and provisional appearance of our knowledge about technology transfer has significant implications, as we shall now see, for the topic of technology controls.

WESTERN TECHNOLOGY CONTROLS: ECONOMIC, FOREIGN POLICY, AND NATIONAL SECURITY OBJECTIVES

Western governments have considered, and in some cases undertaken, the control of West–East technology transfers in the post-war era for three primary reasons: (1) *economic* – to inflict damage on or restrain the Soviet and Eastern European economies; (2) *foreign policy* – to attempt to influence Soviet and Eastern European political behaviour; and (3) *national security* – to restrict the transfer of technology contributing to Soviet and Eastern European military capabilities. Because of the varying assessments of West–East technology transfer addressed in the first half of the chapter, and different economic, political, and security assumptions and considerations in the various Western states, there has been considerable disagreement over these three reasons for technology controls within and among the Western states. In order to understand better the basis and nature of Western technology controls, we will examine thinking and practice in the economic, foreign policy and national security areas. We will find that although various initiatives have been mounted to control West–East technology transfers for all three reasons, the national security rationale is the only one of the three that receives widespread support within the Western alliance today. Accordingly, our discussion will devote more attention to the security reason for Western technology controls.

Economic warfare?

There has always been an element of economic warfare in Western – or at least US – strategic thinking toward the Soviet Union and Eastern Europe in the post-war era. In the initial post-war period, Western trade and technology transfer policy with the Soviet Union and Eastern Europe was mixed. The Western countries, including the USA, exported industrial technology, machines and supplies, but for the most part, embargoed aircraft, ammunition, defence-related technologies, and other exports with military applications. However, as the deterioration of the uneasy wartime alliance with the Soviet Union developed in 1947–8, the USA took the lead in forging a broader policy of economic warfare. For example, the American Chargé d'Affaires in Moscow, George Kennan, cautioned against trade and technological relations which might contribute to the Soviet economic program of 'military industrialization'.[31] In the summer of 1947, the US House of Representatives passed economic warfare legislation directed against the Soviet Union.[32] Key discussions were also taken in the US executive branch in 1947 and 1948. The goal of the USA, as stated in the minutes of one interdepartmental meeting was 'to inflict the greatest economic injury to the USSR and its satellites'.[33] Subsequently the US Congress passed the economic warfare-orientated Export Control Act in 1949, and the USA took the lead in establishing the multilateral export and technology control mechanism, COCOM, in 1949 and 1950.[34] Although many of the European allies resisted, the USA forged a restrictive policy intended to place an embargo on exports, including technology, that might contribute to either military or civilian economic performance. In addition, tariffs were set high, trade and technology transfer facilities and mechanisms were restricted, and credits were discouraged. Overall, Western policy was to deny the Soviet Union and the Eastern European states the benefits of trade and technological relations with the West.

As early as the mid-1950s, however, European resistance to the more restrictive American position resulted in a softening of the policies of economic warfare. Items that were judged of less direct military relevance were dropped from the COCOM embargo list, and West–East trade and technological relations began to expand. The USA was initially reluctant to liberalize its policies, but slowly loosened its controls on trade and technology transfer for economic reasons, joining its allies in more liberal policies in the early 1970s. Although national security controls were maintained, technology controls imposed for reasons of inflicting economic damage on the Soviet Union and Eastern Europe were largely dropped.

However, with the deterioration of detente in the late 1970s, the USA made a number of new proposals that smacked of economic warfare.[35] In response to the Soviet invasion of Afghanistan the USA proposed that COCOM expand its more traditional national security role and broaden its controls to include 'process know-how' in a list of 'defence priority industries' identified by such general terms as 'metallurgy', 'chemicals' and 'aerospace'. The proposal was viewed in many European capitals as politically motivated and intended to inflict damage on the Soviet economy for its invasion of Afghanistan; accordingly, the European governments resisted the US initiatives. In 1981, the Reagan Administration proposed a series of restrictive technology controls on the USSR in response to Soviet complicity in the imposition of martial law upon Poland. Most European governments judged that the Reagan controls went beyond national security concerns, and that America intended, once again, to inflict damage on the Soviet and Polish economies. Although accepting more restrictive technology controls covering military-related transfers, on grounds of national security, Europeans resisted and rejected the economic warfare-orientated proposals. It should be noted that it is sometimes difficult to judge whether US proposals for tighter controls are motivated by economic warfare, foreign policy, or national security reasons. In fact, the reasons are usually intertwined and proposals will often be based, at least to some extent, on two or three of the rationales. What is clear, however, is the strong opinion in Europe resisting the imposition of technology controls intended to impose damage or economic hardship on the Soviet and Eastern European economies.

Some observers criticize the growth of American technology controls in the 1980s as a form of economic and technological warfare imposed not only against the Eastern countries but against America's Western allies as well. Although the US Government attempts to justify the controls on the basis of national security, American efforts to control US-origin technology and subsidiaries abroad through the extra-territorial extension of US law has been vigorously criticized in Europe as a form of American technological imperialism. Some critics see American actions designed to protect and encourage US technological leadership and rest-of-the-world technological dependence; the critics see these controls as having damaging consequences for allies and adversaries alike. Although we will not attempt to explore the validity of these claims here, the question of US technological imperialism will continue to be a contentious issue in both East–West and West–West relations.

Foreign policy

There has also been a strand of Western strategic thinking – again, coming primarily from the USA – that proposes controls on Western technology transfer for reasons of foreign policy. This thinking argues that Western technology transfer can be controlled and decontrolled in ways that can influence Soviet and Eastern European behaviour. Some have encouraged Western governments to take advantage of this perceived political weapon and pursue a policy of economic (or technologic) diplomacy. Such thinking was evident from the early days of the post-war era. Shortly after the war, the American ambassador in Moscow, W. Averell Harriman, urged that the Soviets be shown that US willingness to co-operate in Soviet technological and economic development would depend on their behaviour in international matters.[36] American, and to a lesser extent, Western European policy during the Cold War era was motivated in part by the expectation that Western trade and technology controls could place pressure on Soviet policy-makers, and perhaps encourage them to pursue policies that were more acceptable to the Western nations.

Although economic diplomacy never became a fully recognized and accepted component of Western technology-control policy, it has been at least an implicit element of American thinking throughout the post-war period. In the Cold War period, it took the form of the 'stick', with the implication that America would not loosen its more politically-motivated controls unless the Soviet Union's foreign policy were made more acceptable. During the 1960s and early 1970s, the linkage policy tended towards a more positive 'carrot' approach, intended to induce more co-operative Soviet behaviour. A policy review by the Kennedy Administration, for example, concluded that expanded trade and technology transfer could be used as an enticing inducement, a positive political instrument in East–West relations.[37] The Nixon Administration also pursued positive linkage, believing that a loosening of technology controls could be tied to improvements in Soviet foreign policy behaviour.[38]

The Carter Administration went furthest in articulating and pursuing a policy of economic diplomacy. In his article on 'Trade, technology, and leverage: economic diplomacy', National Security Council official Samuel Huntington cited Presidential Directive No. 28 noting that 'the United States must take advantage of its economic strength and technological superiority to encourage Soviet co-operation in resolving regional conflicts, reducing tensions, and achieving adequately verifiable arms control agreements.'[39] According to this policy, American policy-makers should be prepared to turn technology controls on and off to influence Soviet policy. With the deterioration of detente in the late 1970s, technology control was both expanded and turned 'on'. The 1978 trial of the Soviet dissident Shcharansky, for example, evoked a new US technology control and export licensing requirement for oil and gas exploration and production technology. Then, in response to the Soviet invation of Afghanistan, the USA suspended several US licenes for high technology exports, tightened and expanded technology licensing requirements in COCOM, severely cut back non-technology US–Soviet trade, and reduced the level of funding and other support for the science and technology exchanges.

The Reagan Administration has also controlled the transfer of US technology for reasons of foreign policy. In response to Soviet complicity in the Polish imposition of martial law, the USA invoked a number of controls on technology transfers to the Soviet Union and Poland, including: (1) suspension of issuance or renewal of validated export licenses to the USSR for electronic equipment, computers and other high technology; (2) expansion of the list of oil and gas equipment and technology requiring validated export licenses and suspension of the issuance of such licenses; and (3) a proposal to US allies for further restriction of high-technology exports to Poland. Technology transfer declined further in 1982 when three US–Soviet agreements (space, energy and science and technology) were allowed to lapse in response, once again, to Soviet complicity in the imposition of martial law on Poland. Then the Soviet shooting down of the KAL airliner in September 1983 led to the US cancellation of negotiations aimed at renewing the US–USSR transportation agreement. The US Government estimated that by late 1983 the level of activity in the remaining science and technology agreements was about 20 per cent of the 1979 level.[40]

Despite American initiatives and pressure, the Western European allies have been reluctant, for the most part, to control technology transfer for reasons of foreign policy. As a result, despite sporadic American efforts to use technology controls as an instrument to influence Soviet policy, it has not become an officially recognized part of Western policy.[41]

National security

By far the most important, and the only officially recognized, Western rationale for the control of technology transfer to the Soviet Union and Eastern Europe is based upon national security considerations. Although all of the Western governments recognized the relationship between West–East trade and technology transfer, on the one hand, and national security, on the other, it was the USA that took the dominant role in organizing the post-war system of strategic trade and technology controls.[42] In December, 1947, the US National Security Council determined that 'US national security requires the immediate termination, for an indefinite period, of shipments from the United States to the USSR and its satellites which . . . would contribute to Soviet military potential.'[43] The resulting US controls were said to require 'close scrutiny of all shipments of industrial materials and equipment which have direct or indirect military significance and which are destined for Eastern Europe'.[44] In 1948 President Truman's cabinet recommended a programme to secure parallel export controls by European and Canadian governments. Accordingly, at US initiative in 1949, the multilateral Co-ordinating Committee (COCOM) for strategic trade and technology controls was created. Functioning under an informal agreement (as it still does today), rather than a treaty or executive agreement, COCOM came to include all of the NATO countries (with the exception of Iceland) and Japan. The original COCOM control list was highly restrictive; it included 400 items in 1952 and was increased to about 500 items in the first part of 1954. At that point, however, the more trade-orientated European governments began to resist the more restrictive US-imposed controls; as a result, major relaxations of the embargo took place in late 1954 and 1958.

Due to growing pro-trade interests during this period, Western control policy was transformed from one based upon a sweeping embargo to one based upon selective controls of items of more direct military significance. The change in Western policy was also reflected, albeit rather more belatedly, in US policy where the more restrictive Export Control Act of 1949 was replaced by the Export Administration Act of 1969. The new Act declared that it was US policy to encourage trade and technology transfer with all countries with which it had diplomatic relations, and to restrict only those 'goods and technology which would make a significant contribution to the military potential of any nation or nations which would prove detrimental to the national security of the United States'.[45] Although the USA generally supported an expansion of West–East trade and technology transfer during the ensuing years of the early 1970s, it reacted reluctantly to the relaxation of national security controls. Accordingly, the USA frequently utilized its veto in COCOM to reject proposals to decontrol, or allow exceptions for, items on the national security embargo list. American views were usually accepted, although some governments occasionally ignored their objections (such as in the British sale of the embargoed Spey engines to the Peoples' Republic of China).

In the late 1970s and early 1980s, the US Government began to press for even tighter national security controls.[46] In 1978, the USA proposed that COCOM tighten procedures for the export of computers, clarify its control on software, and tighten controls on automated telephone circuit-switching equipment. In 1980, in response to the Soviet invasion of Afghanistan, President Carter announced a review of US policy that would result in stricter controls. Illustrative of the more restrictive US policy were proposals in COCOM to tighten controls on computers, to require COCOM consultations for the first time on Soviet projects with more than $100 million Western input involving 'process know-how' in 'defence priority industries', and to impose a moratorium on 'exceptions' which allowed the Soviet Union to import items on the embargo list. The inauguration of President Reagan in 1981 further boosted the more restrictive US approach to West–East trade and technology controls. Deeply concerned about the military consequences and strategic implications of what were considered to be lax controls, the Reagan administration proposed a panoply of measures to curtail the eastward flow of militarily-relevant technology. The measures included the strengthening and tightening of US and COCOM controls, expanding controls on 'militarily-critical technologies', cracking down on industrial espionage and other covert collection activities, and reducing Soviet access to Western militarily-relevant, scientific research. President Reagan persuaded the COCOM countries to hold 'high-level' discussions, where, among other things, the US proposed tighter controls on advanced computers, other electronics, fibre optics, semiconductors, and several metallurgical processes. Although many US proposals met with stiff European resistance, the COCOM countries did agree in 1984 to update their controls on computer hardware, software, and telecommunications.

Although there is a solid consensus within the Western alliance supporting the control of West–East technology transfer for reasons of national security, there is continuing controversy on what is to be controlled and how to go about controlling it. Throughout the COCOM experience, for example, the USA has generally attempted to broaden the definitions of what ought to be controlled. Recent examples concerning the proposed control of 'process know-how' in 'defence-related industries', and attempts to control oil and gas technology are illustrative of growing US stringency on technology transfer. In the former, the USA argued that the Western know-how going into the metallurgy, chemical and aerospace industries was making long-term strategic contributions to Soviet military capabilities. In the latter, they argued that items aiding the exploitation and transmission of Soviet oil, gas, and other natural resources, were making strategic contributions because the Soviet leaders would use the resulting hard currency earnings to purchase strategic technology and to finance subversive activities abroad.

Another US effort to broaden the conception of what is strategic – an effort that many now consider to be a failure – concerns the militarily-critical technologies list (MTCL). The 1976 US Defense Science Board Report, called the Bucy report after its chairman Fred Bucy of Texas Instruments, stimulated an immense effort to identify and develop a list of militarily-critical technologies that ought to be controlled.[47] The current MCTL is voluminous; it already fills 17 thick volumes. Many Western governments resist the MCTL approach because they fear that if it is accepted, it will mean an inevitable expansion of COCOM controls. Although some in the US Government now recognize that the effort was far too ambitious, too unfocused and difficult to convert into a specific control list, they will probably continue to push the MTCL approach. A manifestation of this approach that raises considerable controversy between the USA and its European allies concerns more restrictive US licensing requirements on the transfer of critical technologies from the USA to other Western countries. The allies, and many in the USA as well, argue that Western security, and perhaps more significantly, political relations, are damaged rather than enhanced by the expansion of US controls on West–West technology transfer.

The question of 'West–West', or more accurately 'US–West' controls has been a particularly sensitive one in the Western alliance since the pipeline imbroglio of 1982. Although the extraterritorial extension of US controls on the pipeline case was dropped in the fall of 1982, the question of extraterritoriality has been a continuing controversy, particularly as it related to US attempts (1983–5) to renew the Export Administration Act. During this period, the Reagan Administration sought to protect its power to apply US export controls to US subsidiaries, licensees, and other affiliates abroad, and to extend US controls in the form of a US import ban on the products of foreign companies which defy the extraterritorial application of US and multilateral controls.

Another troublesome technology control issue within both the USA and the Western alliance, and likely to be the source of future strategic trade-control controversy, concerns the transfer and control of 'intangible' technology. With the exception of the USA and Germany, members of COCOM do not control what are called intangible transfers of technology. The USA takes the position that controls on strategic technology transfer should cover both tangible transfers, such as computer tapes and discs, and intangible transfers, such as oral exchange and training, or informal technical services. The US Government has even sought to control intangible transfers by limiting foreign access to government-financed, but unclassified, scientific papers (for example, this occurred at a magnetic bubble memory conference during the Carter Administration and at an optics conference during the first term of the Reagan Administration). However, due to negative reaction within the US scientific and academic communities, the US government has backed away from its more restrictive proposals to control scientific exchange.[48] These sorts of controls raise some of the extraordinarily complicated constitutional, political and technical issues with which Western nations will be forced to grapple in future years.

An additional technology control issue, which troubles the alliance, concerns the status of COCOM as an institution and its operational effectiveness. There are some in the USA who would like to see it upgraded to formal treaty status. They are also concerned that it is operating with inadequate staffing, facilities, and resources in a run-down annexe of the US embassy in Paris. These proponents of an upgrading of COCOM – particularly within the US Department of Defense – feel that an organization with such critical relevance to Western security ought to be recognized accordingly, and given considerably more resources to do the job. They would like to see COCOM formalized, and provided with a larger international staff, together with more modern facilities and equipment. Critics of these proposals are concerned that an upgrading of COCOM, particularly efforts to formalize the organization in a treaty, will do more harm than good. They are concerned that raising the issue in parliamentary debates will raise anti-COCOM forces that will be harmful to allied strategic controls. While recognizing that some modernization of COCOM efforts (e.g., word processors) is necessary, they prefer to stick to the organizational arrangement which has worked in the past.

A related controversial question, raised by the USA at both of the 1982 and 1983 high-level COCOM meetings, concerns the establishment of a military subcommittee in COCOM. At Department of Defense insistence, the USA proposed a permanent military subcommittee, to be distinguished from the regular COCOM delegates drawn from the Foreign and Trade Ministries, to provide independent assessments of the military risks of proposed sales. Complaining that the COCOM partners allow insufficient attention from their defence establishments on licensing decisions, and noting that it is often allied commercial ministries or the exporters themselves that are providing technical input to licensing deliberations, the USA argued for more direct ministry of defence participation in the COCOM process. The allies have vigorously opposed the US initiatives. They argue that if there is a lack of relevant military input, which they often dispute, it should be dealt with in the national capitals, and not at the supranational level in Paris. They are also concerned that military representation in COCOM will make it more difficult for the national governments to speak with one voice. And importantly, they are very sceptical of the US Department of Defense's motives in pushing this proposal.[49]

The US Government is also attempting to bring a number of the Asian high-technology producing countries into the COCOM process. In the mid-1980s, US officials were actively involved in bilateral and multilateral talks with officials from these countries to prevent Soviet acquisition of US-controlled technology. Three options were being considered to bring the new high-technology producing countries into the COCOM process: (1) bilateral control arrangements with the USA that might grow into multilateral ones; (2) involving the Asian countries directly in COCOM; (3) an Asian version of COCOM. Although much work needs to be done, US officials see some sort of Asian control system on the horizon. In this regard, Singapore announced in 1985 that it was prepared to co-operate with the US to prevent the transfer of militarily-relevant technologies.[50]

CONCLUSION

The issues of West–East technology transfers and Western controls will continue to be problematic and contentious in the years ahead. Because of uncertainties about the significance and implications of Western transfers to the Soviet Union and Eastern Europe, and further uncertainties about appropriate and effective controls in the present age of an open world economy and massive technological revolution, a variety of views will continue to compete. The West Europeans and Japanese, for understandable political and economic considerations, will tend to emphasize the economic benefits of technology exports and the costs of excessive controls. The USA, on the other hand, will stress the national security costs of Western transfers and the need for restrictive controls. In addition, of course, there are diverse perspectives and competing groups within many of the Western countries ensuring future domestic political controversy on these issues. Under all circumstances, the issues of West–East technology transfer and Western controls will require continued attention in both academic and governmental communities.

7
Technology Flows within Comecon and Channels of Communication

VLADIMIR SOBELL

The intra-Comecon transfer of technology has been a relatively neglected subject. This is not really surprising because despite their emphasis on industrial development the member-countries of the grouping have proved themselves to be technologically weak in relation to the developed Western countries, so that intra-Comecon exchanges of machinery and equipment have come to be regarded, rather contemptuously, as a mutual exchange of inefficiency – a notion far removed from what is normally understood by the transfer of technology. Understandably, it is the West–East transfer of technology (described in the previous chapter) which has received far more attention. It is a far more rewarding territory for economists and students of international (East–West) relations. Yet alongside the increase in West–East trade and technology transfer there have been parallel developments within the Council for Mutual Economic Assistance (CMEA), which complement the former and are stimulated by it. Like the West–East transfer of technology, its intra-CMEA counterpart has a central role to play in the strategies of modernization of the Comecon economies during the transition to 'intensive growth'. It may not involve international diffusion of technology of the same high standards as takes place on the West–East plane, but it provides the CMEA members with the bulk of their needs in medium- or low-level technology not generated domestically. At the same time intra-Comecon co-operation is a vehicle for diffusion within the grouping of high-level technology acquired in the West or developed within Comecon, often by joint efforts.

The aim of this chapter is to consider intra-Comecon transfer of technology in the context of the changing priorities of industrial development involved in the Eastern European transition to the strategy of intensive growth. We shall look first at the issues involved, and subsequently at the role of Comecon institutions in stimulating and promoting this change by means of intra-Comecon co-operation and technology transfer.

COMECON AS A LOW-TECHNOLOGY AREA

I have argued elsewhere that in contrast to market-based integration, the main function of Comecon-type integration is the assurance of stability necessary for effective functioning of the centrally-planned systems.[1] Comecon's collective protection from its unpredictable and inflationary exogenous environment has been based on quite reasonable principles. Given the emergence of a bloc consisting of centrally-planned (Soviet-style) economies, it would probably have been fatal if they failed to close their ranks in the face of this alien environment; but it has resulted in technological isolation from the rest of the world. At first this did not matter too much because the task of the day was to implement rapid industrialization by means of an extensive strategy, which consisted of sheer expansion of capacity and mobilization of labour. The creation of Comecon in 1949 speeded up this process by facilitating the diffusion throughout the grouping of the necessary technology, free of charge. However, much of this technology consisted of well-tried methods and equipment used in the Soviet Union (instrumental in the promotion of Soviet-type industrialization) and was largely obsolete at the point of its diffusion. The participation in Comecon, moreover, implied the adoption of strict central planning, and there is now ample evidence to suggest that this placed powerful brakes on the generation and diffusion of indigenous technological advance (even imported technology is often used with little effect).[2] Thus, despite its important function in promoting industrialization, Comecon was also instrumental in stunting the member-economies' technological progress by promoting essentially conservative industrial strategies and centrally-planned systems unsuitable for the tasks of intensive growth.

The result of this development has been that, in relation to the industrialized West, Comecon has become a region of high cost, low variety and low quality.[3] This in itself would not perhaps be significant were it not for the following factors. First, since comparisons with the West could not be avoided, such results implied a failure of Eastern European industrialization and undermined its legitimacy. Second, and more to the point, with the rapid exhaustion of the potential for extensive growth, this East–West quality (productivity) gap exposed the centrally-planned systems' inability to generate what in this chapter is called *qualitative*, as opposed to *quantitative*, growth. This prompted the governments to pursue various reform programmes but the record of these reforms is generally disappointing and no government has so far succeeded in placing their systems on genuine market-based principles.

In addition to reforms, the CMEA governments have been pursuing other strategies designed to minimize the effects of the systemic inability to generate qualitative growth. The factors of qualitative growth could in the first place be acquired in the form of technology imports from the West and secondly, some gains could be made through increased intra-Comecon co-operation. Imports of machinery and equipment from the West were never completely discontinued, but the CMEA countries were able to benefit more and more as the Cold War receded and COCOM regulations became more liberal. This development coincided with a process of gradual maturing of Comecon industrialization, which objectively demanded a greater dependence on foreign economic relations. However, since the Comecon systems were never fundamentally reformed, the bloc economies could not achieve optimum levels of integration with the world economy (monetary and commodity inconvertibility and chronic shortages of hard currency imposed a definite limit on their ability to do so). But to make up for it there was plenty of scope for progress by means of joint efforts and closer intra-Comecon co-operation. Any gains beyond the margin obtainable through West–East links had to be attained through intra-CMEA projects.

THE TRANSFER OF TECHNOLOGY VS THE TRANSFER OF INEFFICIENCY: A STATIC VIEWPOINT

The discussion of intra-CMEA technology transfer requires a fresh look at the definition of technology transfer because in the customary use of the term (for example, in the context of West–East or West–South trade) we do not normally question whether or not the machinery in question is going to make a positive impact on the technological level of the importer's industry. In the intra-Comecon context, however, a legitimate doubt in this respect does arise as we are dealing with a rather ambiguous paradigm of 'exchange of inefficiency'. Thus, it is imperative that we ascertain the conditions under which the mutual exchanges cease to represent genuine transfer of technology and identify the point at which they become an imported source of inefficiency and an economic liability. In order to do this it is useful first to look closely at the concept of technology transfer, and secondly, to discuss the whole issue against the backdrop of the changing environment in which Eastern European industrialization has been taking place.

In its broadest sense, technology transfer has been defined as follows: 'Technology is transferred from country X to country Y whenever production in country Y is affected by technical information, product or processes from X that have not previously been utilised in Y.'[4] This definition implies that technology is transferred on the basis of a one-off transaction, that is to say that it takes place *only* during the supply of the *first* specimen of a given vintage of machinery or product. But this implies that all subsequent sales of the same products do not constitute transfer of technology, and this clearly is not the case; it would be applicable only in the transfer of technical information or processes (by means of licence) but even that would imply that there is no after-sale feedback or exchange of information. To take into account these more durable aspects of the *process* of technology transfer, another definition by Hanson is perhaps preferable in this context: 'Technology transfer is any process whereby the productivity of resources in one country can be increased by the transmission from other countries of information or of products and processes embodying that information.'[5] On the basis of this definition we can assume that all trade in licences, machinery and industrial end-use products (such as steel semi-products or chemicals) represent a vehicle for technology transfer, but its potential for acting in this capacity is a function of the impact it makes on the productivity levels of the importing country's industry.

In its turn, the effectiveness of this impact is a function of two variables: first it depends on the technological standards of the transferred machinery (processes or industrial goods) and secondly, on the technological (productivity) level of the recipient industry. The higher the former and the lower the latter, the more scope there is for the recipient to make productivity gains and the greater is the potential for trade to act as a vehicle of technology (subject, of course, to the condition that the recipient can assimilate the imported technology).[6] Conversely, the lower the former and the higher the latter, the less potential there is for trade to transfer technology, and one can envisage a situation when the impact acquires a zero or negative value and trade becomes a liability and a drag on the overall productivity levels in the importing country. At this point we can legitimately speak of a situation when such trade objectively ceases to perform the function of the vehicle for technology (even though it may still embody reasonable high technical standards) and becomes a carrier of inefficiency.

THE TRANSFER OF TECHNOLOGY VS TRANSFER OF INEFFICIENCY: A DYNAMIC VIEWPOINT

One of the symptoms of Comecon's slide to the status of a high cost, low-variety and low-quality area has been the cleavage between the indicators of quality (productivity, variety, cost effectiveness) on the one hand, and sheer quantity on the other. For analytical purposes it is useful to distinguish between three different stages of Eastern European industrialization:

Stage 1 represents the initial wave of Soviet-type industrialization which in most countries was completed by the late 1950s. This phase may be referred to as the total extensive strategy (total extensification), when the overriding objectives were the mobilization of unemployed (or underemployed) production factors with the greatest possible speed.

Stage 2 denotes the intermediate phase, the end of which is now looming over the horizon. During this stage the sources of extensive growth began to dry up and the considerations of quality and productivity in the planning process became more important (in proportion to the degree of exhaustion of the extensive factors).

Stage 3 represents the transition to wholly intensively-generated growth (total intensification) during which the factors of extensive (wasteful) growth are nearly exhausted and further growth depends solely on advance in the quality dimension.

In contrast to Stage 1, growth in Stage 3 is generated only by modernization (i.e. the deployment of successive technological generations) and by relevant procedural and organizational (but ideally also systemic) changes. This process of technical restructuring may be regarded as constituting qualitative, as opposed to quantitative, growth: the volume of output of, say, steel may actually decline if an obsolete plant is closed and a new capacity for better quality material is put into operation. Thus, in quantitative terms there may be stagnation or very little growth whilst significant objectives are being accomplished in terms of quantitative growth.

The systemic causes of technological stagnation in the CMEA, noted at the beginning of the chapter, naturally spilled over into intra-bloc trade, with the result that the technical standards of exchanged machinery not only lagged behind Western products, but also behind the changing demands of the East European industries. What may have constituted a significant impact on the productivity levels in Stage 1 in, say, Romania, would have made a diminishing impact as this country (or a given industrial branch) moved to a higher level of development in the latter sections of Stage 2. Unless the machinery in question is continually modernized in proportion to the importing country's (industry's) development, the trade in question will cease to carry its 'productivity load' and might become a liability in Stage 3.

In a market economy the adjustment to changing conditions occurs more or less automatically as firms normally import (and their suppliers are motivated to export) only technology optimally suited to their needs. It is therefore unlikely that a market economy in aggregate would continue to import machinery of stagnating technical standards or unsuitable to a given combination of the available factors of production; in the first place it has the advantage of working with reliable prices, and secondly, individual firms can turn to a competitor who is better able to match their specific requirements. It is unlikely, therefore, that foreign trade transactions would in fact take place unless they served as channels for the transfer of technology.

In the centrally-planned paradigm, however, the lack of reliable price and cost indicators makes it extremely difficult to detect any such subtle shifts in the combination of the production factors. But in any case, given the prevailing seller's market there is generally little an enterprise can do to discontinue unwanted imports. Thus, in intra-Comecon trade the importers constantly run the risk of getting precariously close to the point where the imports cease to carry technology (productivity) improvements and become a burden. This danger has been increasing as these countries approach Stage 3; in Stage 1 and early Stage 2 any machinery imports were a potential source of improved productivity as labour employed in any industrial capacity was more productive than unemployed or underemployed labour. In the late Stages 2 and 3 this was no longer applicable. In order that trade in machinery, equipment and other industrial end-use products can satisfy the attributes of the transfer of technology it must not only be capable of making an impact on the importers' capacity levels, but also of accelerating their factor productivity growth. In other words it must be capable of acting as one of the factors of what has above been referred to as qualitative growth.[7]

COMECON'S 'INVERTED OPTIMUM'

The systemic differences between market and centrally-planned economies, noted in the preceding sections, not only have a definite impact on the decision-making processes in their respective systems, but also influence the evaluation of the outcomes of these decisions. That is to say that any evidence of intensified transfer of technology within Comecon, such as is adduced in the following section, should not be judged by the same criteria which are applied to West–East or West–West technology transfer. The intra-CMEA standards for determining an optimum solution in a given situation are radically different to those normally applied in the West.

The key difference is that the pressure for intensified intra-bloc industrial co-operation, and hence, technology transfer within the grouping, emanates mainly from the need to eliminate technology bottlenecks which have so far either been tolerated, or completely or partially prevented by imports from the West. In the absence of active market tools (prices and profit motive) the CMEA planners have in effect little choice but to fall back on such relatively crude and wasteful scarcity signals. Thus on the intra-Comecon level the planners tend to be moved into action only when the shortages of a given type of equipment become sufficiently acute and economically (or politically) damaging. It is only on reaching this point that things start to move and resources are shifted from the competing allocation to joint development and transfer of technology within CMEA.

In the market environment, on the other hand, the relevant decisions are generally taken well before any supply bottlenecks occur, as firms are motivated by their search for profits and have adequate pricing systems to identify areas of scarcity (any bottlenecks that do occur tend to materialize on the demand side, where they take the form of unemployment). The practical effect of this difference is that while in the East *any* advance in technology is welcome (because it is bound to bring about *some* improvement in the overall productivity level), a Western firm, which struggles to increase its share of the market, will generally be satisfied only if the new technology is *good enough* to beat off the rival products. The market induces companies to reject the average and to strive for technological leadership. In Eastern Europe, on the other hand, leadership is the last thing on the enterprise managers' calculus. If they are concerned with their technology (productivity) levels at all it is mainly in so far as they do not want to end up at the bottom among their 'competitors', thereby incurring the wrath of the authorities.

On this basis it can be concluded that in practice the notion of optimality in the centrally-planned system is not only 'different', but that the former is an 'inverted' version of the latter: the Comecon planners (enterprises) are induced to turn their backs on technological excellence and are preoccupied with steering their enterprises away from the worst; their Western counterparts, on the other hand, are induced not to look back but to strive for leadership.

This principal difference between the Eastern and Western signals for economic decision-making must have some bearing on the evaluation of the results of such decisions. If the Comecon planners take measures that lead to a termination of dependence on imports from the West of a given machine (or material) by joint development and mutual exchange of a substitute, it would be a mistake to denigrate such an effort merely on the grounds that the technology developed in this way fails to match in all technical respects that which was previously imported from the West. What matters to the CMEA planners is not so much the quality (productivity) difference between the CMEA-developed product on the one hand and its Western counterpart on the other, as the fact that the former is able (at least partially) to substitute the latter; the convertible currency thus saved can be more profitably allocated to other use such as the import of technology in an area where the East–West lag is even more pronounced. Correspondingly, the fact that a pressing bottleneck is removed or alleviated is more important than any considerations regarding the product's relative standing in comparison with Western counterparts.

AN ASSESSMENT OF INTRA-CMEA EXCHANGE OF MACHINERY AND EQUIPMENT

Machinery and equipment (i.e. the main vehicle of technology) have always occupied pride of place in CMEA development strategy, and this has been reflected in the grouping's trade. The share of machinery and equipment in total exports rose from 14 per cent in 1950 to over 30 per cent in the late 1970s; its share in total imports rose from about 25 per cent to nearly 40 per cent.

Mutual exports of machinery and equipment have been growing much faster than total intra-Comecon turnover (see table 7.1), and by 1980 the share of machinery and equipment exceeded 40 per cent of total intra-CMEA exports (see table 7.2). Trade in this group of products, moreover, occupies an unusually (in comparison with industrially-developed countries) high share in the total trade of these countries. A study by Fallenbuchl shows that in 1974 the GDR and Czechoslovakia had larger shares of machinery and equipment in total exports than Japan and West Germany – the countries with the largest shares among the Western countries.[8] About 70–75 per cent of Comecon trade in machinery and equipment is directed to the bloc allies. The high shares of machinery and equipment in total Comecon trade and intra-Comecon exports are explained in terms of the priorities given to industrial development, as opposed to consumption of consumer goods; and the apparent high industrial prowess reflected in the higher shares of machinery in total exports compared with those of the Western countries must be seen against the background of Comecon export performance in the West. In the late 1960s CMEA-manufactured exports to the EEC and EFTA amounted on average to about 30 per cent of total exports whereas imports of such goods varied around 80–90 per cent.[9]

In view of the above-mentioned difficulties with identification of genuine transfer of technology (as opposed to the transfer of inefficiency) in intra-CMEA trade, it would be useful to have some method of grading intra-Comecon machinery flows in accordance with their potential as carriers of technology (productivity). Such a gradation would require at least three categories:

(1) grossly sub-standard, obsolescent products (exchange of inefficiency);
(2) goods capable of making some productivity impact but sub-standard in relation to comparable Western products;
(3) machinery and equipment (and industrial semi-products) approximating or exceeding the prevailing Western standards.

Table 7.1 Indices of mutual turnover and exports of machinery and equipment*

	1960	1970	1980
Mutual trade turnover	3.2	7.8	25.4
Mutual exports of machinery and equipment	4.9	14.5	50.8

* 1950=1
Source: V. Moiseenko and E. Egoshin (CMEA Secretariat), *Ekonomicheskoe sotrudnichestvo stran chlenov SEV*, 1982, No. 10, p. 24.

Table 7.2 Shares of machinery and equipment in total exports to CMEA

	1970	1980
Bulgaria	33.7	54.3
Czechoslovakia	59.8	63.2
GDR	59.8	64.0
Hungary	45.2	47.1
Poland	52.3	59.9
Romania	28.3	40.3
USSR	21.8	23.5
CMEA total	39.7	41.1

Source: as table 7.1.

In the absence of such statistics it might be helpful to consider as a surrogate the data relating to the selection of machinery and equipment for the projects linked with intra-Comecon co-operation and specialization. This can be done on the following grounds: as regards category (1) we may be reasonably certain that such equipment would not be included in co-operation and specialization agreements (whatever we may think of the Eastern European planners it is hard to imagine that they would deliberately plan an exchange of products which would actually decrease the productivity levels of their industries – although in individual cases such errors could conceivably occur due to sheer incompetence). On the other hand, products falling into category (3) and above-average products of category (2) do on the whole stand a reasonable chance of being co-opted into the mainstream of intra-Comecon co-operation and specialization exchanges.

For reasons discussed below, the process of selection for such projects accelerated in the 1970s. Exports of 'specialized'[10] engineering products grew nearly twice as fast as total mutual supplies in this group of tradeables and reached a share of 38 per cent in total engineering exports in 1982[11] (the share declined slightly the following year). Although much of this growth can be attributed to the sheer increase in the number of specialization agreements, and hence nomenclature coverage, the fact remains that in the early 1980s, well over a third of intra-CMEA exports of engineering was sanctioned as comprising technology with a potential of making a significant impact on the present and future productivity levels in CMEA. The shares of specialized exports in selected branches of engineering are shown in table 7.3.

Table 7.3 Shares of specialized exports in total exports by groups of engineering products

	1975	1980
Machine tools	25.0	46.4
Machine presses	11.2	43.2
Electrical equipment	10.6	35.6
Lifting and transport equipment	46.2	61.1
Textile industry equipment	25.2	53.1
Chemical industry equipment	25.5	34.0
Road transport and construction equipment	40.5	49.8
Communications equipment	15.3	24.9
Ball-bearings	20.6	87.3
Tractors	39.8	44.5
Agricultural machinery	31.9	44.6
Lorries and maintenance equipment	3.4	43.2
Ships, shipping and port equipment	81.6	81.3

Source: as table 7.1, p. 142.

Obviously one can be sceptical about the meaningfulness of such figures. To start with, one would expect at least one third of Eastern European exports to be 'above average' in any case (to make up for the 'below average' third at the other end of the scale). Secondly, the fact of selection need not necessarily imply that the machinery in question is the best in Comecon.[12] Apart from the general scepticism regarding the claims that the machinery included in intra-Comecon specialization constitutes the best that the member-countries can offer, a determined critic might suggest that even if such claims were borne out in reality, we would still not be justified in treating such data as evidence of intensified transfer of technology (as opposed to inefficiency) within Comecon because, to put it simply, where would such a sudden surge of technology come from given that the CMEA is a 'low-technology area'?

Several points must be made to answer this criticism. As shown above, the Comecon countries now have no choice but to improve their technological performance if they are to implement a decisive shift towards qualitative growth (intensification); they have little choice because the factors for extensive growth are simply not available any more. Moreover, as will be argued below, the USSR is now determined to extract greater value (in terms of productivity and quality) from its CMEA partners in return for supplies of fuels and raw materials. The intra-bloc industrial co-operation and specialization schemes may be perceived as the tools by which this is realized.

Admittedly it is not so much the Comecon countries' determination which is doubted, as the ability of their industries to deliver the goods. In this connection it must be conceded that the spectacular rise in the number of specialization agreements and in the share of 'specialized trade' itself does not really give grounds for claiming that a significant improvement in intra-Comecon exchange of productivity is underway. However, as any student of the communist economic systems knows, in order to achieve any improvement, much 'administrative heat' has to be first generated in the form of directives, publicity campaigns and taut planning targets. The actual results may not be as good as envisaged, and everybody involved (the planners and the enterprises) knows it, but *some* improvement is as a rule in the end attained. At the minimum it is plausible to claim that the selection of products for co-operation and specialization has the long-term effect of phasing out exports of grossly obsolescent machinery, whilst enhancing mutual exports of products of above-average standards, often approximating (incorporating) Western technology. This will not eliminate the dependence on the imports of Western technology, but it may reduce it, or ensure that in the future such dependence will be less than it would otherwise have been. In view of Comecon's 'inverted' notion of optimum this might just be regarded as a satisfactory outcome.

Table 7.4 gives some data on the absolute levels of mutual supplies in several categories of products as well as an indication of the share of Comecon-origin machinery and equipment in total imports of the Comecon countries in 1980 – the high point of the expansion of 'specialized' exports. The ranking of these categories according to the rising share occupied by non-Comecon imports is revealing: it shows that all of the groups of equipment where the contribution made by Western equipment was one-third or more (from rank 6 downwards) consist of relatively complicated machinery with relatively high productivity impact and R and D content. These groups also represent the strategically-important capital goods which determine the overall technological standards of the industrial production base and of the final products. Co-operation with the West in these branches is especially rewarding, if not essential. On the other hand, the groups of machinery and equipment more readily supplied from Comecon (ranks 1–6) consist of products which, from a technological point of view, are relatively undemanding.

THE PRESSURES FOR INTENSIFIED TRANSFER OF TECHNOLOGY

Although Comecon has been effectively isolated from the rest of the world economy, it has continued to use world market prices as the basis for intra-bloc prices. This, however, has led to a rather spurious application of the former. Intra-CMEA prices tended to be higher than world market prices, but the divergence was highest in machinery and equipment and not so high in energy and raw materials. This divergence was extraordinary since it contradicted the

Table 7.4 Intra-Comecon supplies of machinery and equipment in 1980

Rank[a]	Type of equipment	Unit of measurement	Mutual supplies	Share of CMEA goods in total imports (%)
1=	Buses	Thousand units	16.5	97
1=	Lorries	Thousand units	65.7	97
2	Tractors	Thousand units	39.8	96
3	Passenger cars	Thousand units	319.0	95
4	Freight railway wagons	Thousand units	6.8	93
5	Lifting and transport equipment	Million rubles	1205	82
6	Energy and electrical equipment	Million rubles	1771	66
7	Chemical, paper and cellulose industrial equipment	Million rubles	5855	65
8	Machine tools	Thousand units	28.4	64
9	Ships and equipment	Million rubles	922	63
10	Instruments, laboratory and medical equipment	Million rubles	1128	60

[a] ranking according to the declining share of CMEA-origin goods in total imports (column 5).
Source: as table 7.1, p. 142.

actual supply–demand pressures within the grouping. Thus, in practice, prices became less important. Instead, the tradeables came to be divided into the 'soft' and 'hard' categories, where the hardest commodities tend to be fuels and essential raw materials, while the softest products are the overpriced and unwanted manufactures. In this way a system of 'unequal exchange' evolved, which penalized the exporters of hard commodities and favoured the exporters of manufactures. By the mid-1960s there was only one net exporter of hard commodities – the USSR – and all the remaining European members of Comecon (including Romania and Poland) became net importers. The Soviet Union was willing to subsidize[13] its 'world socialist market', but, as is well known, exerted pressure on its members to invest in Soviet extractive industries and transport infrastructure. The Eastern Europeans complied, but some economists pointed out that joint investment was no solution to the underlying problem, which was rooted in the low technical standard of their manufactures; the real solution lay in upgrading the technological levels of Eastern European industries.

Despite such claims, joint investment in the USSR reached unprecedented proportions in the 1970s, although the alternative policy of 'hardening' the Eastern European export structures was also pursued. No major new investment projects were started in the 1981–5 Plan, but the Comecon Summit Meeting held in June 1984 in Moscow (and the subsequent Session of the Council held in Havana) prepared the ground for another round of joint investment in energy and raw materials, planned for the next decade.[14]

Writing amidst preparation for the Summit, the Soviet economist Bogomolov (the semi-official spokesman on Comecon affairs) wrote that in the forthcoming decades the centre of gravity of the CMEA integration should move away from co-operation in the provision of ever-increasing volumes of raw materials, to co-operation in technological restructuring, which would lead to better processing of the available resources.[15] There have been some signs that the latter aspect of integration will be pushed more vigorously than in the past. The Summit Meeting decided that a joint 'Comprehensive Programme for Scientific and Technical Progress' for a period of 15 to 20 years should be prepared for approval by the 1985 Session of the Council. In addition, each country was asked to prepare an energy and raw materials conservation programme.

At the Summit the USSR made it clear that the maintenance of the present levels of deliveries of energy would be contingent not only on contributions to investment, but also on Eastern European success in the implementation of agreed industrial restructuring, aimed at the 'hardening' of exports to the USSR. The Summit communique stated that:

[in order] to create economic conditions for ensuring the continuation of deliveries from the USSR of a number of types of raw materials and energy carriers, such as to satisfy the import requirements in volumes determined on the basis of the co-ordination of plans and long-term accords, the interested CMEA member states will gradually and consistently develop, within a framework of an agreed policy, their structures of production and exports and will carry out the necessary measures for this in the field of capital investments, reconstruction, and rationalization of their industry with the aim of supplying the Soviet Union with the products it needs, in particular, foodstuffs and manufactured consumer goods, some types of construction materials, and machines and equipment of a high quality and of a world technical level.

These developments indicate that technological backwardness is rapidly becoming less tolerable. The requirements of domestic 'intensification' have conspired with pressures of exogenous origin and there is now no other way out but to reform and modernize.

JOINT PURSUIT OF MODERNIZATION AS A FORM OF TECHNOLOGY TRANSFER

Since industrial co-operation and specialization lead to the creation of international production systems, the technology involved becomes more readily diffusible thoughout the participating economies. Thus there occurs a transition to a new mode of technology transfer: as opposed to the 'traditional' trade, in which technology is supplied in its entirety by the donor to the recipient, under the new arrangement it tends to be *jointly developed and pooled*. Although Comecon is some way behind the industrialized West as far as the intensity and depth of this process is concerned, the trend pointing in this direction is unmistakable. Since this subject has been extensively covered elsewhere,[16] I shall concentrate here only on the strategically most important and most recent aspects of co-operation.

The main pillar of the modernization strategy is the planned expansion of electronics-related industries and the belated pursuit of 'electronization' both of production processes and the final products. Electronization is expected in the first place to contribute directly to the process of 'lightening' Comecon's excessively energy-intensive industrial structures, and secondly, mechanization (robotization) and computerization should make the industries more efficient. Thirdly, the application of microprocessors in engineering products and consumer durables is expected to upgrade the technical standards of such products and hence to contribute to their 'hardening'. Joint efforts in this strategically important field began in the 1970s, but produced relatively modest results: essentially the organization of the production of a largely obsolete range of computers on the basis of specialization. Mutual exchanges in this group of equipment grew rapidly, and the 1982 Session of the Council sanctioned the first multilateral agreement for co-operation in the production of industrial robots.

The progress of electronization in Comecon has been hampered by the relatively low level of electronics components produced within the grouping and especially by the slow transition to microprocessor circuits. In order to help to eliminate this bottleneck, the 25th Session of the Council held in 1981 concluded the General Agreement on multilateral co-operation in the creation of a unified range of electronic components, the production of specialized lines for electronics, and the production of semiconductor and other high purity materials. The implementation of this programme is expected to last in the first instance up to 1990. In the following year, the 26th Session signed an agreement on co-operation in the production and application of microprocessors. The circuits were grouped into three *Module* groups, depending on their technical properties, and production was divided among the participants. Mutual exchanges of circuits soon commenced.[17]

Dependence on the West, however, is bound to remain considerable. In 1984 it was estimated that only about 55–60 per cent of all specialized materials (in terms of the number of items) were wholly covered by intra-Comecon supplies.[18] Although the Eastern European sources are, as usual, reticent about the precise role of East–West co-operation, preferring to stress the value of intra-Comecon links, there can be little doubt that in many instances the importation of licences and production facilities from the West represents the optimum and widely-resorted-to strategy for the implementation of electronization.[19] Indeed, the progress of intra-bloc co-operation in this field was made possible by the imports of Intel microprocessor technology by the main participants (GDR, USSR, CSSR, Poland and Hungary).[20] In this sense it can be said that intra-Comecon co-operation amounts to little more than co-ordination on the bloc-wide level of the acquisition of Western technology.

DIRECT LINKS AS CHANNELS FOR TECHNOLOGY DIFFUSION

The transition to increasingly internationalized production and R and D, where technology is jointly developed and pooled rather than passed from the donor to the recipient, requires the evolution of direct links among the participating units. Between market economies these links are developed and maintained by the multinational companies whose operation is facilitated by the creation of common markets and financial convertibility. In Comecon, however, the absence of financial convertibility, and the predominance of strict central planning, demand that such direct links are attained by different means. Briefly speaking, the lack of propensity for forging direct linkages is compensated for by artificial, administrative means, essentially consisting of intergovernmental agreements. It is possible to differentiate between several types of such agreements according to their complexity and qualitative features. In the first instance one can distinguish between the relatively simple bilateral agreements (often involving co-operation between individual enterprises from the participating countries) and the more complex general multilateral agreements: the latter stipulate the basic aims and purposes of a given project while the detailed obligations are subsequently determined on a bilateral basis. Secondly, with regard to their qualitative attributes it is possible to make a distinction between agreements (bilateral or multilateral) directly stipulating the terms of co-operation on the one hand, and agreements leading to the setting up of a permanent international body or organization such as the standing commissions, interstate committees, IEOs, or R and D co-ordinating centres (CoC) on the other. The latter are more effective, since they lead to the creation of a network of international organizations whose function is akin to Western multinationals. These organizations remain subject to the strictly centralized conditions governing the intra-CMEA relations, but since they are imbued with a degree of decision-making autonomy, they are in the position to form more flexible international ties than would otherwise have been possible.

The preferred form of linkage-making in the early years of Comecon consisted of the creation of standing commissions for co-operation in a given industry or sector. These grew fastest in the mid-1950s and early 1960s and at present there are 21 standing commissions.[21] In the 1970s there arose a need for more devolved and operative forms of co-operation (standing commissions operate at the level of industrial or sectoral ministries) in response to the need to form internationalized production systems and to diffuse the available technology more efficiently. Thus, there occurred a proliferation of international organizations of various shapes and sizes determined by the specific function they are designed to fulfil. There have been many attempts to bring some order into this microcosm of Comecon-allied bodies,[22] but there are grounds for suspecting that no classification can be entirely satisfactory and exhaustive because the size, administrative structure, legal status and the nature of their linkages with the other CMEA bodies or national agencies are all determined by their immediate operational needs rather than some grand design conceived in all its details at the top. Hence, in chart 7.1 they are depicted as organizations supportive of the Comecon structure proper under their generic heading of *Functional organizations*. At the time of writing there are 34 multilateral, and several bilateral organizations of this kind, but since this number includes those with purely financial, co-ordinating, service or marketing functions, table 7.5 lists only those international organizations that play

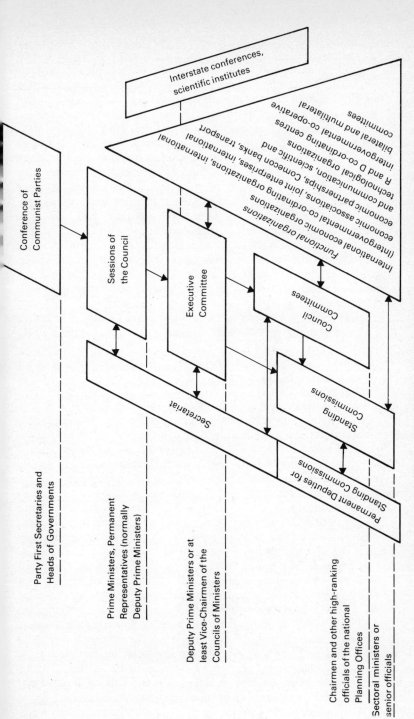

CHART 7.1 *The structure of co-operation and specialization institutions in COMECON*

Source: V. Sobell, *The Red Market: Industrial Co-operation and Specialisation in Comecon*, Aldershot, Gower, 1984, p. 19.

some part in the intra-Comecon development or diffusion of technology.

Table 7.5 does not, however, include the fastest growing (in numerical terms) form of functional organization: the scientific and technical co-ordinating centres (CoCs). These organizations are of the utmost importance in the present context because they extend directly the 'common pool' of scientific and technical knowledge and because they represent relatively efficient and bureaucracy-free (by Eastern European standards) forms of international linkage. CoCs are based on the principle that a leading research institute in a given field is entrusted with the task of co-ordinating Comecon-sponsored research on selected high priority problems. This form of co-operation requires little additional administration and should be capable (at least in principle) of producing faster and cheaper solutions, because individual research tasks are allocated to various sister institutes throughout the grouping. The number of participants in a CoC varies between 20 and 50, depending on the importance of the project, but there have been cases of giant CoCs, such as that housed at the L. Karpov Physical Chemistry Institute of the USSR, which co-ordinates the activities of more than 100 agencies. This CoC was established in 1971 and charged with the implementation of a research programme into anti-corrosion protection of metals.

Table 7.6 gives some information on the dynamics and country-distribution of CoCs. It may be seen that in 1982 well over half were located in the USSR.

FUTURE PROSPECTS

It might be held that the present analysis takes an unjustifiably optimistic view of the prospects of industrial modernization in Eastern Europe accomplished by Comecon's own efforts. Indeed, a healthy dose of scepticism is certainly called for. If the national R and D systems of the member-states have so far failed to provide the necessary technological impetus for their economies, why should the extension of those systems into a bloc-wide level fare any better?

An assessment of the significance and effectiveness of intra-CMEA technological co-operation must take into account the following considerations. First, as is argued elsewhere in this paper, progress in Comecon should not be judged by the same demanding criteria as those applicable in the West. In other words, a little progress (which certainly cannot be denied) is in this case better than no progress at all. Secondly, it is one of the consequences of Comecon's 'inverted' incentive for action that (as long as the present system remains in place) the Eastern European countries are bound to remain technological followers rather than leaders. With a few exceptions, the CMEA R and D establishments have so far proved to be a little more than mechanisms for adapting Western technology to more modest indigenous conditions. The emerging bloc-wide system should, therefore, be viewed in a similar light: essentially it amounts to management on the international plane of the adaptation and diffusion, within the grouping, of technology acquired outside. In general, one can suggest that the 'higher' the level of technology in question, the more relevant is this conclusion. That Comecon is apparently 'getting its act together' in this way is

Table 7.5 International organizations participating in intra-CMEA diffusion of technology.

(A)	*Intergovernmental organizations*	(separate legal status)
	Organization for Co-operation in the Ball-Bearings Industry	(Also known as OSPP or Interpodshipnik)
	Intermetal	(ferrous metallurgy)
	Agromash	(fruit- and vegetable-growing machinery)
	Interkhim	(small-bulk chemicals)
	International Commission for Co-operation in Computer Technology	
	Interelektro	(electrical equipment)
	Intersputnik	(satellite communication)
	International Centre for Scientific and Technical Information	
(B)	*International economic organizations*	(no separate legal status)
	Interatominstrument	(nuclear industry instrumentation)
	Interatomenergo	(nuclear energy equipment)
	Intertekstilmash	(textile industry machinery and equipment)
	Interkhimvolokno	(chemical fibres equipment)
	Interetalonpribor	(measurement instruments)
	Interelektrotest	(high-voltage equipment)
	Intervodochistka	(water purification equipment)
	Internefteprodukt	(oil-processing additives and catalysts)
	Medunion	(supplies of medical equipment)
(C)	*Bilateral organizations*	
	Interkomponent	(electrical and electronics equipment; Poland and Hungary)
	Elektroinstrument	(electric motors and other equipment; USSR and Bulgaria)
	Intransmash	(industrial transport and warehouse equipment; Hungary and Bulgaria)
	Assofoto	(photochemical materials; GDR and USSR)
	Domokhim	(household chemicals; GDR and USSR)
(D)	*International research institutions*	
	International Laboratory for Powerful Magnetic Fields and Low Temperatures	
	International Mathematical Centre of S. Banach for Training of Mathematicians	
	International Centre of the Academies of Sciences for Training in the Problems of Heat and Mass Exchange	
	International Centre of the Academies of Sciences for Training in the Field of Electronic Microscopy	
	Temporary International Collective for Joint Testing of the VVER Nuclear Reactors	
	Co-ordinating Committee for Scientific Instruments and Automation of Scientific Research	
	Interprogramma (bilateral Soviet–Bulgarian organization for computerization of management systems)	
	Joint Institute for Nuclear Research (Dubna near Moscow)	

Sources: compiled from V. V. Vorotnikov and D. A. Lebin (eds), *Mezhdunarodnye ekonomicheskie i nauchno-tekhnicheskie organizatsii stran chlenov SEV*, Moscow, 1980; and V. Morozov (CMEA Secretariat), *Ekonomicheskoe sotrudnichestvo stran chlenov SEV*, 1984, No. 5, p. 63.

not because it expects to transfer itself into a 'second Japan'; it is rather a more modest response to the fact that the disorganization characteristic of previous

Table 7.6 Number of scientific and technical co-ordinating centres

	1971	1975	1982
Bulgaria	—	2	2
Czechoslovakia	4	6	5
GDR	1	4	4
Hungary	—	2	5
Poland	2	5	5
Romania	—	3	5
USSR	11	25	37
Total	18	47	63

Source: *Ekonomicheskoe sotrudnichestvo stran chlenov SEV*, 1975, No. 5, p. 98 (for 1971 and 1975); and 1982, No. 7, p. 50 (for 1982).

joint efforts was too costly and proved to be ineffective in achieving its aims.

The early 1980s were marked by increased hostility between East and West. Comecon was forced to close ranks in the face of Western embargo and vulnerability to what in the East was perceived as 'technological blackmail' on the part of the US Government. This latter consideration certainly influenced the decisions taken at the CMEA Summit of 1984 and at the Havana Session of the Council that followed it. It is possible that this lurch towards isolation will turn out to be transient[23] and that the Comecon-wide R and D system will continue to be developed not just in order to galvanize bloc resources in the face of Western hostility, but for more positive reasons that have more to do with the intrinsic value of science and technology as a determinant of economic performance. In either case, it is reasonable to expect intra-CMEA technological linkages to expand and perform the modest but vital function for which they were designed.

8
Soviet Planning Reforms from Andropov to Gorbachev

DAVID A. DYKER

Except in the crisis area of agriculture, the last few years of the Brezhnev era were characterized at best by drift, at worst by immobility. From the economic, if not from the political point of view, this is surprising. The 1965 attempt at general reform of the economic planning mechanism, which marked the beginning of that era, initiated a brief spurt in Soviet economic performance, but by the mid-1970s the downward trend in growth rates, already evident during the Khrushchev period, had emphatically reasserted itself (see table 8.1). That

Table 8.1 (Average) annual rates of growth of Soviet national income

Year(s)	Growth rate (%)	Year	Growth rate (%)
1961–5	6.6	1979	2.2
1966–70	7.8	1980	3.9
1971–5	5.7	1981	3.2
1976	5.9	1982	3.5
1977	4.5	1983	3.1
1978	5.1	1984	2.6

Sources: various editions of Economic Commission for Europe, *Economic Survey of Europe*, New York, United Nations; Plan Fulfilment Report in *Ekonomicheskaya Gazeta*, 1984, No. 5, pp. 5–8.

trend in turn reflected mediocre productivity performance in conditions where labour and some raw material resources were becoming increasingly scarce and expensive. The traditional Soviet planning system had laid primary stress on increasing production through increasing inputs massively: essentially, the political mobilization of resources. By the 1970s, realistic prospects for further economic growth could only be based on a strategy of increasing the efficiency with which inputs were used, of increasing productivity all round. That is why successive stabs at economic reform have tried to make the system more cost-conscious, more quality-conscious, and technologically more dynamic (the major theme of this book). The 1965 planning reform sought to achieve these objectives by marrying profit and sales-based success-indicators (the traditional indicator had been simply gross output) into what remained a highly centralized system, with the planning organs in Moscow continuing to tell enterprises what to produce, and where to deliver it to. Because it is impossible to do all the necessary calculations implicit in that degree of centralization, plans continued to be poorly co-ordinated; as a result, chronic supply uncertainty persisted as a major feature of the industrial scene. The price system was reformed, with the introduction of a capital charge for the first time, but the central planners continued to fix the great majority of prices. Moreover, the maintenance of centralization in production planning meant that in any case there was little scope for enterprise directors to react to prices in the way in which they do in a market economy. For the same reason, profit did not work well as a success-indicator.

Whilst the 'marketizers' amongst the economists and planners took this as an indication of the need to go further along the road towards some kind of market socialism, to give profit-based indicators a chance to work properly, more conservative elements took it as proof that market-orientated indicators had no place in a socialist system. The latter may have been supported in their anti-market stance by 'perfect computationist' economists, who saw increasing scope for a rationalization of central planning through computers.

A decree on the planning system published in 1979 gave no clear guidelines on fundamental dilemmas such as this.[1] It was, in any case, never fully implemented. Certainly the perfect computationists seemed to be in the ascendency, with the marketizers for the time being under a cloud. But the political apparatus seemed primarily intent on sitting tight and waiting for the old man to die. When that finally happened in 1982, and with the accession of Yuri Andropov to the General Secretaryship of the Communist Party, there was a sharp change in the policy-making atmosphere. A series of hard-hitting articles on the problems of Soviet planning appeared in the press, Andropov himself set out a number of new or refurbished policy principles, and July 1983 saw publication of a decree setting up an industrial planning experiment. By August 1983, the General Secretary's illness seemed to herald the end of this vigorous new impetus in planning policy. Andropov's death had none of the galvanizing effect of Brezhnev's death. But Chernenko did not scrap Andropov's planning experiment, though he may have soft-pedalled somewhat. With Gorbachev now in command, the experiment, recently extended in coverage, remains the framework within which industrial policy alternatives are projected. In trying to analyse post-Brezhnev trends in planning, then, we have to go back to that series of initiatives that came out of Andropov's first six months, and see how they have been developed and transformed since then.

If we consider the policy slogans directly attributable to Andropov and the Politburo, together with the ideas (with their varying degrees of official blessing) of economists and planners, we can begin to put together a complex of inter-related diagnoses and proposals which begin with the proposition that, in a situation of widespread labour shortage, labour productivity is the key to economic growth. More specifically, it is argued that there is substantial scope for increasing productivity through better labour discipline, less absenteeism, etc., and that planned redundancy and redeployment on the basis of job conflation and job inter-changeability offers wide possibilities for slimming down enterprise labour forces. Closely linked to this is the notion that Soviet industrial structure needs to be fundamentally reorganized to bring it into line with contemporary notions of rational specialization. This relates particularly to the engineering industry.

On the motivational side, 'self-management' (the Russians use the same word as the Yugoslavs) is predicated as a proper and necessary part of shop-floor relationships. More specifically, the brigade in industry and construction, and the 'link' in agriculture, are endorsed as key forms of grass-roots workforce organization, involving a degree of self-management and autonomy in the field of financial incentives. More self-management should mean less bureaucracy, which fits with the proposition that the industrial ministries need to be cut down to size, in terms of both staffing and powers over associations and enterprises. By the same token, the degree of the autonomy of the latter should in some sense increase, opening up a vista of hierarchies of autonomous associations, enterprises and brigades. The key to greater sophistication in production is seen in terms of closer monitoring and specifically quality- and new technology-orientated bonus arrangements, the key to fuller satisfaction of customers' needs in terms of stricter enforcement of contracts. Attacking the problem of the quality of the planning system itself, Andropov and his associates argued that planning should be based more on 'stable norms' for incentive funds and investment funds: in other words it should become more parametric. At the same time, the general level of tautness, manifest in pressure for maximum planned growth, should increase. Let us look at some of these points in greater detail.

Labour productivity and over-manning

Labour practices had become very easy-going under Brezhnev. Piece-work tariffs were slack, and workers resisted attempts to tighten up on them by operating their own 'bogeys', or unofficial quotas, just as do Western workers, and by voting with their feet: always an easy option in a fully-employed economy.[2] An in-depth study of construction workers from 1980 reveals a picture of veteran workers almost working when they pleased.[3] But there were already signs of a desire to control this state of affairs before Brezhnev's death: indeed, the decree that gave Andropov the legislative basis for his labour discipline campaign was passed in early 1980.[4] The new element introduced by Andropov was a much increased readiness to punish recalcitrant workers by reducing their labour coefficient, *koeffitsient trudovogo uchastiya* (KTU), which determines their effective wage rate.

Over-manning has always been a marked characteristic of the Soviet industrial scene, and in the pre-reform days enterprise directors had very limited powers of dismissal. Under the rubric of the Shchekino system, first introduced in 1969, some enterprises had been permitted to make workers redundant, subject to finding them another job with the help of the local authority, and to use part of the monies thus economized to pay wage supplements to the survivors. Reports on the operation of the system in the 1970s were uniformly favourable, but while coverage did increase steadily, probably only about 10 per cent of the industrial labour-force was working under full Shchekino conditions in 1980.[5] The reasons for this sluggishness in extension seem to have been essentially four-fold. First, the Government was nervous about the danger of substantial pockets of unemployment emerging. Secondly, superior bodies showed themselves rather too ready to claw back enterprise shares of wages-fund economies. Thirdly, enterprise managers, used to the demands of quarterly output or sales targets and the vagaries of the Soviet industrial supply system, showed reluctance to part with the surplus labour they were holding 'just in case'. More generally, low average productivity in many enterprises, especially in engineering, reflected sub-optimal scale and low technological levels in auxiliary operations, rather than dead wood on the main production line. Over a period of decades chronic supply uncertainty had induced a pattern whereby components and instrument manufacture, the production of castings and the like, had become concentrated in enterprises of sub-optimal size, or in 'dwarf-workshops' within large enterprises. Thus, Soviet engineering enterprises are typically all-purpose organizations, with a main production line surrounded by auxiliary shops making nuts and bolts and other basic components. This is why they are much bigger than engineering enterprises in the West, and it is also why they report much lower average levels of productivity than in the West; on main production activities the productivity gap is quite small.[6] Attempts to change this pattern have come up against the powerful vested interest of the industrial ministries in the existing pattern.

Self-management (samoupravlenie)

Self-management has been something of a taboo word in the Soviet Union in the past, while work-team autonomy has been at times in fashion, at times very definitely out of fashion. In his key article on the teaching of Karl Marx, Andropov posited that

the system functions and is perfected through the process of continually finding new forms and methods of developing democracy, extending the economic rights and opportunities of the working man on the production floor, and in all dimensions of socio-political activity. . . . This is real socialist self-management of the people. . . .[7]

Here work-team autonomy is firmly identified with self-management, and while Politburo member Aliev stated categorically in his speech on the 1983 Law on Working Collectives that 'our concept of self-management differs fundamentally from its anarcho-syndicalist interpretation',[8] there can be no doubt that the new law has provided a basis for extension and consolidation of brigade and link systems. What this means in principle is that work-teams, rather than just following instructions received day by day from a hierarchical superior, receive an output or sales target for a period of, say, three months, an overall limit on wages and bonuses backed up by an appropriate cash advance, and the necessary material supplies. Beyond that they are on their own.

We should not exaggerate the importance of the brigade in industry and construction. Autonomy is only meaningful, and profitable, if it is possible to count on all supplies coming through, which brings us right back to perhaps *the* fundamental problem of the Soviet economy: supply uncertainty. In any case, in the Andropov reading, brigade autonomy seems to get inextricably mixed up with the favourite themes of labour discipline and 'getting the finger out', as 'working collectives . . . elaborate and adopt counter plans ("voluntary" extra plans) confirm measures to raise labour productivity and socialist competition arrangements . . . apply social incentive measures . . . and impose penalties for infringements of labour discipline'.[9] In agriculture, links are less vulnerable to industrial supply breakdowns, since they are organized primarily to *tend* crops which have already been sown, manured and sprayed, or to look after and feed up livestock. Under the rubric of the 'collective contract', there has been substantial development of agricultural link and brigade autonomy since the death of Brezhnev, though the actual content of the autonomy varies considerably from one farm to another.

The planning bureaucracy

Ministries and ministers have always tended to be whipping boys in the Soviet Union. In 1957 Khrushchev found it economically and politically convenient to blame them for various inefficiencies leading to falling capital productivity, and he did, indeed, abolish them altogether. They were brought back in 1965 because the regional economic councils that replaced them were even more heavily indicted for wasteful use of investment resources than the industrial branch ministries had been. The fact is that ministries and regional economic councils alike have been victims of the system. Because it is the ministries who have to see to the implementation of plans, which are often patchily thought out and badly co-ordinated, they are almost bound to cut corners and break rules in the name of plan fulfilment. We have already seen how they are forced to distort the industrial structure in order to ensure supplies of key components. More immediately, they often chop and change association and enterprise plans, as they juggle with the production capacity available to them, and wheel and deal on the grey markets to obtain scarce supplies. It is hardly surprising that a leadership as concerned with 'stable norms' (see discussion below) and socialist legality should look askance at this kind of thing. From a quite different, and much more radical angle, the Novosibirsk sociologist Zaslavskaya, in her 'secret' seminar paper, pin-pointed the ministerial system as the focus of a top-heavy, over-bureaucratized system which had outlived its usefulness as an engine of crude growth maximization.[10]

Planning for quality and the quality of planning

The 1965 planning reform had sought to make producers more sensitive to quality and specification by introducing profit and sales as key 'success-indicators'. We saw that this did not work because profit cannot operate as a success-indicator unless there are some significant market elements in the planning system, while as long as client enterprises are not allowed to shop around when consignments fail to satisfy, sales is in practice not very different from the notorious old gross output success-indicator. Brezhnev's 1979 planning decree expressed a new confidence about the scope for generalized computerized quality-control systems, and proposed that the details of inter-enterprise contracts should increasingly be settled by the enterprises themselves – but with the centre still deciding what the basic contractual links should be. This essentially legalistic approach to the quality/specification problem was closely echoed in Andropov's favourite trinity of labour discipline, production discipline and planning discipline; a decree of August 1983 specified that production of sub-standard work should be treated, and punished, as a straightforward infringement of labour discipline.[11] But how is it possible to *enforce* planning discipline? Faced with a colossal task in trying to plan more than 15,000 commodity groups, and often under pressure from the political authorities for maximum short-term output growth, Soviet planners have tended to work on the 'ratchet principle', namely, on the basis of a mark-up on what was achieved in the previous period. However valuable as an easily applied rule of thumb, this principle has the disadvantage that it encourages enterprises with 'hidden reserves' to keep them hidden, and to avoid the introduction of new technologies, which may depress output performance in the short run. The contrasting principle of stable norms, which has figured in every planning decree since 1965, seeks to establish targets and bonus parameters for at least five years ahead, so that enterprises need feel no fear that short-term gains in performance and bonuses may be quickly confiscated as the planners catch up with them. Since the death of Brezhnev there have been dozens of articles in the Soviet press endorsing the parametric (*normativnyi*) principle. Nevertheless, we must be sceptical about any commitment to stable norms not backed up by a decentralization substantial enough to ease the planners' burden to a significant degree. Without such a decentralization, hard-pressed Gosplan officials will continue to use the ratchet principle as a condition of survival, and that means arbitrary adjustment of targets from year to year. Even more palpably, there is a fundamental contradiction between stable norms and taut planning. Yet Chernenko insisted at the April 1984 Plenum of the Party Central Committee that 'we cannot possibly get by without a further increase in the level of tautness of our work in the economy'.[12]

These, then, are the major themes of post-Brezhnev economic policy-making, and many of them figure prominently in the 1983 decree setting up an experimental industrial planning system.

THE INDUSTRIAL PLANNING EXPERIMENT

The main elements in the industrial planning experiment, first introduced operationally in selected sectors in January 1984, and extended in coverage in January 1985, are as follows:

(1) As a general principle, the role of production associations and enterprises in the drafting of plans should increase. At the same time assessment of plan fulfilment by the centre is to become more rigorous. This implies a reduced role for ministries and their sub-units in plan drafting *and* in the process of 'adjustment', whereby traditionally intermediate planning bodies have guaranteed achievement of the aggregate plan, while ensuring that

most subordinate organizations would report plan fulfilment. Plan implementation should become more parametric.
(2) Key success-indicators are sales/deliveries according to contracts for all producing units, and where appropriate 'development of science and technology', quality, growth of labour productivity, and cost reduction or increase in profit. Profit continues to be the source of finance for the bonus fund, but will apparently lose its special role in relation to fund-forming coefficients.
(3) Stable norms for wages fund, incentive funds, and so forth, are to be established on a five-year basis.
(4) No bonuses must be paid to managerial workers unless sales/deliveries plans are met in full.
(5) Autonomous production association and enterprise control over decentralized investment, financed from profits through the production development fund, is to be re-established (this had been a feature of the original Kosygin reform of 1965). Production units are also to enjoy greater freedom in financing 'technical re-equipment' (*tekhnicheskoe perevooruzhenie*) from amortization allowances and credit.
(6) Production associations and enterprises are to be permitted to use monies from the ministry-level Unified Fund for the Development of Science and Technology to finance autonomous R and D work, and to compensate for increased costs in the period of assimilation of new products.
(7) Producing units are to be allowed more autonomy in deciding the allocation of the socio-cultural and housing fund, financed through a stable rate of deduction from profits.
(8) Management is to be given greater independence in the use of bonus funds and wages fund economies accruing through job rationalization.
(9) Budgetary rules are changed so that production units can retain a larger proportion of profit on a regular basis. This implies that the various incentive funds are to grow in size as well as independence, though it may also imply an increase in the proportion of *centralized* investment financed from retained profits.[13]

On the face of it, there is some genuine increase in the freedom of manoeuvre of the enterprise. The number of planning indicators imposed by the ministry has been reduced from 30 to 10.[14] On the other hand, the constellation of key, fund-forming indicators seems, if anything, to increase in complexity under the experiment. Thus,

planned deductions into the material incentive fund are calculated on the basis of rates of reduction of costs. For every percentage point of cost reduction this year the planned base year size of the material incentive fund increases by 5 per cent. In addition, a stable norm for deductions from accounting profit is established. In case of non-fulfilment of the cost-reduction plan the planned level of profit will not be achieved, which will reduce the rate of deductions into incentive funds.[15]

What are the implications of all this; and what happens when quality, productivity and innovation, not to mention deliveries, are brought into the picture?

Enterprises on the experiment certainly are being allowed to keep and spend, without higher approval, substantially greater proportions of profits. One enterprise reports that its production development fund grew from 2.3 million rubles in 1983 to 3.3 million in 1984, providing a basis for the implementation of 19 investment projects. In addition, 1.5 million rubles' worth of re-equipment and reconstruction investment, including the installation of automatic systems, is being carried on on the basis of a State Bank loan.[16] But the director of another enterprise complains that with a fund of 1.2 million rubles in 1984, they were only able to secure material supplies and equipment to the value of 790,000 rubles. What will happen in 1985, when the production development fund grows to 1.6 million rubles?[17] The manager of a third enterprise points out that it is virtually impossible for small and medium-size enterprises to get a construction organization to build houses for them – so what do they do with their socio-cultural and housing fund?[18] In fact, the Soviet economy went through precisely these difficulties in the late 1960s and early 1970s, and for the same reason: it is no use decentralizing on the purely financial side, if there is no concurrent decentralization on the material supply side. Brezhnev's answer to the problem in the mid-1970s was to claw back decentralized funds. If the current experiment is going to fare better it will surely have to provide something more radical on the supply planning side than the Soviet Union has yet seen.

In general, the new system is extremely conservative on the supply issue. True to the Andropov creed, quality is to be improved in the first instance through computerized quality-control systems, and workers who turn out bad products are being fined up to one-third of their average wage.[19] Still on the discipline theme, management really do appear to be losing all their bonuses when contracts are not fully met.[20] But what happens if supply hiccups originate from inconsistent plans and/or poor work by material supply depots? Under the experimental system the enterprise management always carries the can, although 'any industrial leader will agree that just about every production problem usually arises from unbalanced, unstable plans'.[21] Because the experiment is only an experiment, it could hardly be expected to affect the overall level of centralization of the Soviet planning system, and therefore the general problem of imbalance in plans, to a great degree. But there are in any case remarkably few 'free trade' elements in the experimental system as a whole.

Having said that, we must also note that some light industrial organizations in the Baltic republics and Belorussia have been conceded greatly extended freedom in the supply area under the rubric of the industrial planning experiment. Right from the start, enterprises under the Lithuanian Ministry for Local Industry have been allowed substantial freedom to negotiate patterns of deliveries to retail outlets.[22] In an article published in early 1985 the director of the 'Komsomolka' textile association in Minsk describes an arrangement whereby supplying enterprises are encouraged to fulfil their contracts with the association through direct contributions to the bonus funds of those enterprises *by the association*. For all practical purposes, that means that the association is being allowed to negotiate its own materials prices with suppliers. The 'Komsomolka' director also confirms that the association negotiates prices for high-quality goods directly with the retail network.[23] There is some evidence, then, of an 'experiment within an experiment' which takes a much more radical approach to the supply problem.

Beyond that, the supply side of the experiment seems to repeat the same old stories of an over-centralized, over-bureaucratized system that we have been reading about in the Soviet press for 20 years. One director complains that Gossnab, the State Supply Committee, insists on sending him materials rendered redundant by technical progress; but that if he tries to sell them off directly he faces a fine.[24] Another asks how on earth he is supposed to plan for contract fulfilment in advance, when the delivery plan for his crane factory has not even been provisionally confirmed. The draft production plan is supposed to be elaborated six months in advance, but with just a few months to go before the start of the new planning year only 75 per cent of orders (*naryad-zakaz*) for cranes for 1985 had been confirmed. Looking forward to the five-year perspective, the enterprise has no idea what the balance of orders is going to be for its two main production lines – cranes and mining equipment. So it has no basis for planning how to spend its new-found wealth earmarked for technical re-equipment.[25]

Enterprises operating under the experiment have been understandably concerned that contracts imposed on them should be for manageable quantities, and Gossnab made a ruling at the beginning of 1984 that local supply organs should issue allocation certificates on enterprises involved in the experiment only for quantities big enough to be containerized – for smaller quantities the supply base itself should act as wholesaler. In practice, supply depots are ignoring the ruling.[26] 'The Main Supply Office of the Ministry for the Machine-Tool Industry is forcing us to dispatch production to its numerous customers in penny numbers. ... We are getting 12,000 dispatch notes a year. Why turn the factory into a corner shop?'[27] To make matters worse, ensuring payment for every delivery is just as important as making every delivery, since the precise formulation of the key experimental indicator is in terms of 'sales according to contracts'. 'We have to send "pushers" to all corners of the country ... [to try to get the money]. Sometimes we make deliveries totalling 50 per cent above plan and still barely make our aggregate sales target.'[28]

To repeat, all these themes are familiar. But in the old days of gross output or aggregate sales targets, enterprises could get away with fudging the details of plans. Now that enterprise managements stand to lose all their bonuses if all contracts (12,000 of them?) are not met in full, the clumsiness of the supply-planning system takes on a completely new significance.

But it would be wrong to dismiss the experiment as a non-starter. The brigade system is being used as a vehicle for the introduction, through the labour coefficient system, of more precise payment-by-results schemes,[29] and of a very tough approach indeed to absenteeism and lateness for work.[30] Perhaps most important within the province of operating directly on the labour productivity variable, have been the strikingly high rates of planned redundancy in some enterprises on the experiment – with the principle of voluntary formation of brigades ensuring that anyone who finds himself left out has only himself to blame. The minister for the energy equipment industry tells us that the wages funds for design workers in selected organizations in his ministry have been cut by some 18 per cent through redundancies since the ministry embarked on the experiment.[31] In the 'Ukrelektromash' association, 450 out of a total labour-force of some 5,500 are to be made 'conditionally' redundant by the end of 1985.[32] The force of the 'conditionally' is probably that the men will be found new jobs, either within the association, or nearby. But, of course, the question arises of what kind of macro-economic tremors this sort of thing might produce if practised on a national scale. And at the micro-economic level too, we can take note of the experience of the Khar'kov electrical engineering factory, where the work-force has to devote considerable amounts of time to straightening, stretching and welding sub-standard materials – because they are not allowed to send them back to the supplier.[33] As long as the experiment does nothing to sort out such basic supply problems we must expect enterprise managers to remain deeply ambivalent towards schemes for slimming down labour-forces. The workers may remain equally ambivalent, as they ask themselves what they stand to gain. As we saw, the basic principle of these redundancy packages is that a substantial proportion of the funds saved should go in wage supplements to the remaining workers. In at least one enterprise included in the experiment it has been made quite clear that in order to keep these special supplements, workers will have to earn them. In particular, they stand to lose the lot if they fall down on delivery commitments.[34] Thus the experiment treats workers as harshly as managers in relation to failures which individuals or groups may be helpless to prevent.

In a speech made in early 1984, Prime Minister Tikhonov described the industrial planning experiment as 'a broad, perhaps the broadest ever, search for and elaboration of the most effective principles and methods of planned socialist management of the economy'.[35] That may be something of an overstatement. However laudable the aims of the experiment, it fails altogether, in its primary form, to grapple with the characteristic feature of the Soviet planning system which lies behind most of the problems: over-centralization. For that reason, the attempt to parametricize planning will not work, because at the given level of work-load, planners simply do not have enough information to calculate the parameters. Moreover, supply uncertainty will remain a key feature of the Soviet industrial scene, with all that that entails in terms of inhibiting attempts to establish high-productivity orientations in industrial structure and manning patterns. If extended industry-wide, the principal experiment would probably increase the general level of organizational efficiency by a few percentage points, as long as the principle of no bonuses unless all contracts are fulfilled did not result in a complete seizing-up of the whole economy. But it is inconceivable that the current stage should represent the end of the road for the industrial planning reform process. Rather, the experiment presents Mr Gorbachev with a framework embodying a range of possible policies for the future. The 'no-change' option is the most obvious one, but the firm, if cautious step in the direction of marketization taken in the area of light industry could have far-reaching consequences if generalized. It is, *inter alia*, precisely the sort of thing that is required to resolve the contradictions of those fundamental provisions relating to autonomous enterprise funds. There is no sign at the time of writing of any major reorientation in the industrial planning experiment under the new leadership. Gorbachev was clearly keeping all his options open when he said, in April 1985, that 'while strengthening the main features of centralised planning, we propose to broaden further the rights of enterprises, to introduced genuine business accounting, and on this basis to increase responsibility for, and material interest in, the final results of work, at the level of the collective and of the individual worker. That is the purpose of the economic experiment'.[36] Thus the General Secretary is free at any time to take any dimension of the experiment and develop it into a new economic policy orientation.

CURRENT POLICIES ON RESEARCH AND DEVELOPMENT

The thinking behind the R and D orientated elements in the industrial planning experiment is obvious enough. Major new technologies can be implemented centrally through direct instructions to enterprises. Smaller-scale innovations can be left to the enterprises themselves, with increased scope for internal and external finance provided. Supply problems apart, the main difficulty with the latter proposal lies in the inevitable ambivalence of managerial attitudes. As long as directors have to put the main priority on fulfilling delivery plans, quarter by quarter, they may be less than enthusiastic about any innovation which threatens to disrupt continuous production, if only briefly. As far as the former is concerned, we can only pose the question: what is being done to speed up the traditionally lethargic Soviet R and D establishment in order to ensure that the 'major new technologies' are not obsolescent before they are brought on-stream?

In September 1983 a decree was published 'On Measures to Accelerate Scientific-Technical Progress in the National Economy'.[37] It was a disappointingly vacuous document, which used many words to berate everyone from the Academy of Sciences and the State Committee for Science and Technology downwards for failing to implement a unified science and technology policy; no precise explanation was offered as to how the unified policy was in fact to be achieved. The decree posited the creation of giant scientific-production associations, following a policy line of the 1970s, which had not been an outstanding success; reduction in managerial bonuses by at least 25 per cent where research and development assignments had not been met; and a 30 per cent price supplement for new products embodying new technology, plus a 30 per cent price drop for obsolescent products. It is not clear that the threat of a 25 per cent bonus cut would significantly alter the priorities of a manager who knows that failure to fulfil basic sales targets may mean no bonuses at all and trouble with his superiors. The ±30 per cent price rule was a good idea, but unfortunately it was left to the State Committee for Prices to decide how to apply it, with all the suffocating bureaucracy that that entailed.

There are also special provisions in the decree to allow ministries to sanction extra bonus payments above the normal legal limit for managers who introduce new technologies. This may provide enough incentive to overcome managers' general distaste for innovation, though it runs an obvious risk of encouraging 'simulation'. It is in any case unlikely to provide sufficient incentive for managers to go for innovations which are bad for sales/output trends even in the medium term, because, for example, they permit one component to do the work of two. We can illustrate this point very simply with a quote from the manager of a textile factory on the planning experiment.

'Hard' five-year plans, strict long-term plans, and with the planned levels of production always going up, are, if one may say it, a hindrance to textile workers. For example, today knitted dresses are in demand; by tomorrow cotton dresses might be all the rage, and no one will want to buy knitted. The labour costs on these articles are identical, but the difference in price is of the order of 4:1. In the chase to fulfil aggregate value plan indicators, which determine wages, textile workers are forced to turn out old-fashioned, sometimes expensive items, thus producing, so to speak, only for the warehouse shelf.[38]

No wonder Soviet machines, even good ones, tend to be too heavy![39] And no wonder the Soviets are having particular difficulties in trying to develop a micro-chip industry. All the failings of the Soviet R and D establishment seem to be summed up in the story of the Agat microcomputer, which finally came on-stream in mid-1984 – at least a year late.[40] Thus, technology lags which are serious enough in any case are compounded by planning failures. Excessive lead-times represent one of the symptoms of the complex of problems which affects R and D in the Soviet Union. As we shall see now, the cause of effective assimilation of new technologies may suffer at least as much at the stage of design and construction of specific enterprises.

THE INVESTMENT PLANNING DECREE

With design/construction lead-times two to three times what they normally are in the West, and capital costs often escalating sharply, the Soviet Union has been ill-equipped to cope with the investment needs of sectors with rapidly changing technologies. By the early 1980s, the incremental capital-output ratio (ICOR) in the Soviet economy was around 7:1.[41] Part of the background to these problems lies in the traditional planning regime in design and construction, under which performance was assessed and bonuses paid out largely in relation to physical output: of drawings, of holes dug, of bricks laid, and so forth. There was no incentive to aim for speedy completion. Often, building enterprises found it financially advantageous to start a new project rather than to finish an old one.

Attempts to reform design and construction along the lines of the Kosygin industrial reform of 1965 failed because it was found to be impossible to develop a meaningful sales-type indicator for application to large-scale, one-off projects. Experiments conducted in Belorussia in the late 1970s sought to put all the emphasis, in terms of finance and bonuses, on commissioning of completed projects (*vvod v deistvie*). This was a qualified success, though there were problems in working out financial flows and bonus entitlements over a construction time of perhaps five or more years. The solution adopted was to employ the concept of marketable output (*tovarnaya produktsiya*) as an indicator of putative stages of a completed project. This indicator is supposed ultimately to be based on the operationalization certificate of the completed project, so that the two indicators of marketable output and commissioning should in the end come to the same thing. In practice, this has not happened, so that marketable output suffers from many of the weaknesses of the old volume of construction indicator, which had calculated tranches of work done, without regard to project completion.

Throughout the 1970s the system of design and construction organizations continued to manifest other perennial weaknesses such as organizational fragmentation and lack of technological sharpness. Meanwhile the ministries proved incorrigible in their typically bureaucratic vice of always wanting to get as many projects going as possible so as to strengthen their negotiating position in divisions of future cakes. So they, too, must take a big share of the blame for the continued crisis of *raspylenie sredstv* – 'excessive investment spread' – with far more projects on the go than the design and construction sectors could possibly cope with.

By the publication, in June 1984, of the first comprehensive decree on the fixed capital formation process for a number of years, the Chernenko leadership signalled its intention of having another go at this complex of problems.[42] The operational core of the decree laid down that by 1 January 1985 all design and estimate documentation for projects carrying over into the next Five-Year Plan would be reviewed with the aim of cutting out unnecessary elements, postponing projects of secondary importance, and ensuring that embodied technologies were up-to-date. This proposal seems to involve at least three major difficulties. First, the work is to be devolved to the ministries, which will add greatly to their already heavy planning load, though an extension of the industrial planning experiment might help to even things up. More fundamental, however, is the question of whether the ministries, who may be more interested in bureaucratic jostling than in cost-efficiency, can be trusted to do the job properly. Secondly, as long as design organizations, like construction enterprises, continue to be rewarded for quantity rather than quality of work, can *they* be trusted to shift the priority to cheapness and technological dynamism? If not, then the work-load imposed on central and ministerial planners will be that much greater. Thirdly, the new system of estimate norms and prices introduced for construction and investment on 1 January 1984 is in any case wreaking havoc with the estimates situation. As of April 1984, only 65 per cent of estimates for projects due to be commissioned in that year had been revised.[43] With prices in the sector generally rising under the new regime, and with some estimates revised and some not, it is difficult for the ministries to make coherent plans for getting estimates down. Once again, the implications for planners' work-loads are horrific.

The decree also includes a number of provisions for improving the technical quality of investment planning. Planned levels of investment are to be better balanced with plans for investment finance, supply of building materials and equipment for installation, and the production capacity of the building industry. This is indeed a very laudable aim, but in the absence of any general decentralization and rationalization of the planning system as a whole, it looks like a non-starter. The Soviets are clearly pinning their hopes for improved co-ordination in investment on the introduction of their computerized 'Unified System of Planning for Capital Investment' (ESPKS). Introduction of the system, which started in 1982, is not scheduled to finish till 1986, and we should probably expect delays. There are also question marks over the capacity of Soviet, or indeed any, computer systems to grapple with such complex problems. Most important of all, as one Soviet investment specialist argues,[44] the first condition of improving co-ordination in Soviet investment is to scrap half the projects under construction at the present time. That would go far beyond the timid rationalizations proposed by the investment decree, and would stand in greater need of political toughness than of fancy computers.

More concrete is the proposal that commissioning should be the key planning indicator. Still, no solution to the problem of how to use it as a basis for a quantitative, work-in-progress indicator is offered. Housing and amenities investment is now to be planned in strict conjunction with production investment, and commissioned when the latter is commissioned. Certainly no one could quarrel with the desirability of such co-ordination, and there is a clear attempt here to back up one of the priorities of the industrial planning experiment. But if we think back to the kinds of co-ordinational problems which have already arisen in connection with enterprise housing investment, we must be sceptical as to whether an area where co-ordination has been *particularly* poor in the past is really going to improve rapidly.

Feasibility studies (TEOs) are to be reintroduced as an independent element in project planning for large-scale and complicated projects. There has always been a special problem of low quality in this kind of design work in the Soviet Union, and in 1981 it was decided to abolish TEOs as such, and integrate their content into an upgraded system of regional and sectoral development and location schemes. That clearly did not work, so they are going back to the old system. The ultimate reasons for the existence of this problem are essentially two-fold. First, the ministries are not interested in locational rationality. Secondly, design organizations are badly under-paid for feasibility study work. The new system of estimate norms and prices may solve the second problem. The fact that it is left to the ministries to organize the TEOs, as it was left to them to organize the development and location schemes, suggests that the first problem will continue to impinge for some time yet.

Still on the theme of a more integrated approach to investment complexes, the turnkey (*pod klyuch*) system, whereby construction organizations are commissioned to build and deliver a fully operational production complex, is to be systematically introduced in selected regions, including Belorussia. Once again, we have to say that, in the absence of general and substantial measures to improve the supply situation, this is unlikely to be widely developed, though it may work well for a limited number of top-priority projects. But it is taken very seriously by the Soviets, and there have been a number of back-up decrees.

Finally, provisions appear to be made for a degree of decentralized contracting between enterprises, construction and design organizations, in relation to reconstruction and re-equipment projects. This clearly ties in with one of the most important elements in the industrial planning experiment, and the theme is to some extent followed up in a recent article by a number of authors, including the distinguished investment specialist, R. M. Merkin, which maps out (whether as a reality or a possibility is not clear) a series of 'broadly-based economic experiments' for the investment and construction sector.[45] Under experimental conditions, a building organization which ordered more of a particular material than envisaged in the plan, or went back on an order placed earlier, would have to pay a 5 per cent fine. Unsatisfactory supplies from industrial enterprises would be payed for at the rate of only 80–85 per cent of the list price. This seems to suggest that: (1) building enterprises should be allowed a good deal of freedom on the details of supply arrangements, within the overall framework of the plan; (2) in effect, construction organizations should be allowed to buy in extra inputs if they are prepared to pay a premium; (3) the same organizations should be allowed some freedom to bargain over just how much a below-par consignment of supplies is worth.

But there are other elements in this article which seem to pull in a different direction from the investment decree. On success-indicators, there is scepticism about the advisability of going for rapid commissioning 'at any price', and a proposal that the sole key indicator should be 'final economic results': achievement of planned production trends over the lifetime of the project. (This, of course, immediately raises the question of how you base actual payment on 'final economic results' in an economy where the average project takes from seven to ten years to complete!) Again, the construction experiments would be based on large-scale combined design-construction associations, whereas in fact the investment decree confirms the traditional trust as the basic organizational unit in the building industry. Thus, the relationship between the decree and the experiments in the investment and construction sphere is problematic, and it is quite impossible to gauge to what extent the latter should be taken as a development of the former, or to what extent indeed it may represent some kind of 'minority report'. It is extremely odd that in the issue of the journal *Economics of Construction*, which carried the article, there is also an editorial on the investment decree, but with no cross-referencing between the two pieces whatsoever. Nothing has appeared since Mr Gorbachev's accession to clarify the point, but the most positive evaluation of the investment situation might be along the lines of our evaluation of the industrial planning experiment. Recent enactments, plus experiments proposed or implemented, give the General Secretary a spectrum of policy choices ranging from extreme caution through to pragmatic radicalism. As yet he has shown no signs of picking one or the other.

CONCLUSIONS

In the light of these details, what can we make of the style and tone of Soviet economic planning and policy-making at the present time? We can start by picking up and generalizing the last point made in the previous section. There seems to be an almost obsessive concern to keep all possible options open, and we should not forget that Prime Minister Tikhonov announced in early 1984 that the Communist Party had decided to press forward with a 'Programme for the Complex Perfecting of the Management Mechanism'[46] – which could prove the answer to everything, or give everyone an excuse to do absolutely nothing for years. The experience of the Chernenko and early Gorbachev periods does, indeed, suggest that there is a great deal of sheer inertia to be overcome by any Soviet leader bent on a genuine reform. More specifically, we must recognize that the Communist Party apparatus has a vested interest in the old system, which cannot function at all without political trouble-shooters. It is perhaps a reflection of the innate conservatism of the industrial planning experiment that it seemed, under Chernenko at least, to signal a reaffirmation of the role of the Party apparatus in the economic sphere. If contracts are to be the key, and if the supply system remains unreformed, then it stands to reason that it will be the *apparatchik* and the Party committee which will in the last analysis have to see that those contracts are kept to. In one article from early 1984 a provincial Party secretary cites with applause the history of a brigade leader and Party group organizer who came in to help out, at another enterprise, when a machine broke down and plan and contract fulfilment was threatened by the slowness of the official maintenance channels. For days, we are told, he fiddled about until finally he found the fault and repaired it.[47]

In fact, Party activists have been performing this trouble-shooting role since the days of the first Five-Year Plan, and there can be no doubt that it was often very effective in the old gross-output days. Whether it can handle the greater subtleties of contractual details and advanced technologies is another matter. In addition, Party apparatus men may find it difficult to combine the trouble-shooting role with that of 'guardian of socialist legality', equally stressed in recent articles. After all, the hero who fixed the machine must have needed some components, and must have obtained them outside the official supply system, which caused all the trouble in the first place. But we should not be *too* condescending about this revival of a more traditionally Bolshevik approach to problem-solving. If nothing else, the Party apparatus can surely do much to consolidate what is perhaps the crudest element in the post-Brezhnev package: simply pressurizing people to work harder. Our provincial Party secretary, for instance, holds up the shining example of a collective farmer who worked round the clock to thresh 7,500 centners of grain single-handed (well, almost!) in a way reminiscent of the Stakhanovite movement of the 1930s. The problem with that kind of approach is that it can only work for so long, though with leaders changing so rapidly at the moment the 'new broom' effect may be expected to extend its life somewhat. In the longer term, however, workers will surely go back to playing safe, just as apparatus men will go back to wheeling and dealing, because ultimately that is the only way to survive in the Soviet economy. Nothing definite has appeared since the death of Chernenko on the role of the Party in the economy. Gorbachev is of impeccable apparatus background, and got the job because the apparatus trusted him. But the option of sacrificing the Party apparatus for the sake of a decentralizing, technocratic reform must represent a dream – or a nightmare – for most of the people in the top Soviet elite.

Can the Party professionals look for any comfort from the perfect computationists among the economists and computer scientists? Certainly a world-class Soviet computer industry would do something for Soviet economic planning, though it would do much more for Soviet production processes. The problem is how to set up the economic environment in such a way as to tap the potential of computerized procedures. Here we can do no better than quote the words of Academician N. Fedorenko, director of TsEMI, the Central Economic–Mathematical Institute: 'The main obstacles to the assimilation of mathematical–economic models and methods are the deficiencies in the economic mechanism.'[48] More specifically, Soviet managers are always in favour of computerization when it helps them to raise output, and always against it when it helps the planners to improve their assessments of production capacity, to uncover 'hidden reserves'. But in the light of our earlier discussions of the essential nature of the Soviet planning system, can we really be surprised by that? As leading Soviet mathematical economists have been saying since the time of Khrushchev, economic reform and rationalization of planning procedures, through computers and any other technology which is available, are simply different aspects of the same necessary process.

Finally, just how serious is the Soviet economic crisis? Officially reported growth rates are still in the 3 per cent range, and while we should certainly be sceptical about the 'quality' of that reported growth, we should be equally circumspect about the recalculations which come up with a 'real' rate of growth of around zero. To put the issue in perspective, the Soviet Central Statistical Administration claimed an average rate of growth of Net Material Product of 4.2 per cent during 1976–80, while the CIA calculate a corresponding rate, in GNP terms, of 2.7 per cent.[49] As long as aggregate growth rates stay in the 2–3 per cent range, and as long as economic recovery in the West remains hesitant, the pressure for more radical reform may continue to meet with resistance. On the other hand, a straight comparison of the USSR's 2.6 per cent and the USA's 6.7 per cent in 1984 must give the Kremlin some food for thought. In any case, as we have seen in a number of the other chapters in this book, there are specific technological problems and lags which may impinge very sharply on development prospects for specific sectors of the Soviet economy, particularly the defence industry. It is, of course, arguable whether a planning system which still insists on judging its industrial executives on the criterion of obedience (real or apparent), and which may take up to ten years to complete major investment projects, is capable of solving these specific problems. Thus at the *micro*-economic level the pressures for further reform may be that much more difficult to resist.

9
Prospects for the Soviet Economy
DANIEL L. BOND

Previous chapters of this book have dealt primarily with various aspects of Soviet science and technology. In this concluding chapter we will focus on developments in the overall Soviet economy, and look at its prospects for the remainder of the 1980s.

Technological progress is a term often used to identify that portion of growth which cannot be directly linked with the increases in the observable factors of production: land, labour, capital or materials. Western and Soviet economists place heavy emphasis on the role of technological progress in maintaining the rate of growth of an economy. And advances in science and technology are often seen as the major source of technological progress. Thus it is reasonable to seek in a country's economic performance signs of the success or failure of its science and technology. Another link between the performance of a nation's economy and its progress in science and technology is the role that the economy plays in providing the resources for the latter. It is obviously easier to secure the resources for supporting the costs of science and technology when the economy is healthy than when it is not.

RECENT SOVIET GROWTH TRENDS

The performance of the Soviet economy over the last decade raises questions concerning both the contribution that science and technology is making to the economy, and the ability of the economy to support increased efforts to advance science and technology in the future. There has been a gradual, long-term slowing of Soviet economic growth in the post-war period, as is indicated by the decline in growth rates depicted in figure 9.1. Estimates of the growth of aggregate and sectoral productivity indicate that there has been a corresponding slow-down in technological progress in the Soviet economy.[1] Some of the underlying causes of this slow-down are:

(1) the declining numbers of new workers joining the labour force, and the diminishing supply of labourers capable of moving out of agriculture and into other sectors;
(2) the ageing of the capital stock – due to excessively low rates of replacement – with a corresponding increase in resources required for capital repair and maintenance;
(3) the depletion of natural resources in the more accessible regions, resulting

in rising combined costs of mineral extraction and transport;
(4) a decline in opportunities for quick growth through 'catching-up' as the relative backwardness of many sectors in the economy has been reduced;
(5) a long-term decline in labour and managerial discipline and morale, in part due to the lack of adequate improvement in material incentives.

In the latter part of the 1970s the Soviet economy entered a period of more serious growth deceleration. This appears to have been the result of the combined effects of several factors:

(1) sectoral imbalances which were the result of the excessively rapid expansion of some sectors of heavy industry (energy, machine-building, chemicals) relative to the growth of the underlying industrial infrastructure and sectors supplying materials inputs to heavy industry (such as rail transportation and ferrous metallurgy);
(2) a reduction in the overall rate of growth of capital investment, combined with increased allocations of investments required by the energy and agricultural sectors, which delayed capacity expansion in many critical sectors;
(3) the loss of output and general disruptions resulting from unusually adverse weather and a series of poor harvests.

The longer term trends and sectoral imbalances combined to create numerous bottlenecks in the Soviet economy which brought growth almost to a standstill by the end of the 1970s.[2] Since this explanation of the recent Soviet growth slow-down is the key to our view of the economy's future prospects, some concrete examples of these bottlenecks are warranted.

Soviet railway freight traffic expanded rapidly during the 1960s and 1970s, but growth has fallen off since 1975. Full capacity utilization was reached on key stretches of the rail line and at the major stations during the mid-1970s. Many factors contributed to this, but mainly it reflects the decision made in the late 1960s not to invest as much in expansion of the railway system. As a result, by the late 1970s, rail transport appeared no longer capable of meeting all the demands that were being placed on it, and the resulting disruptions in the materials supply system were a bottleneck holding back output in other sectors of the economy. Finally, the extreme winter of 1979 appears to have touched off a 'cascading collapse' of the components of the rail system from which it has still not fully recovered.[3]

A similar situation arose in the Soviet construction industry. Here too, rates of output and productivity dropped in the 1970s, in part due to underinvestment in this area relative to the demands being placed on it. As a result there was a serious mismatch between the planned demand for construction services and the real capabilities of the construction sector. The resulting build-up of unfinished construction contributed to the break in growth trends in the economy that occurred in the late 1970s.[4]

Although it is difficult to measure the degree of imbalance in the overall economy, this can be suggested by examining the relative rates of growth of its various components. It is not expected that output of all sectors of an economy

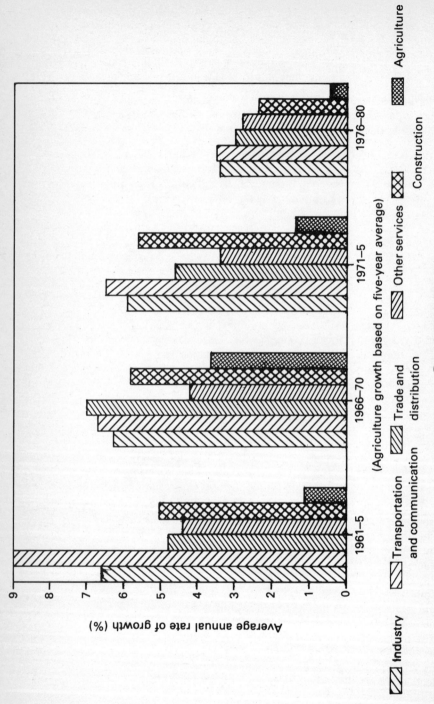

FIGURE 9.1 *Growth of GNP by sector*

will increase at the same rate over time. Variations can be necessitated by changing composition of final demand, foreign trade or inter-industry requirements. However, if because of faulty planning or other reasons, some sectors grow more rapidly than those supporting them with services or material inputs, it is likely that a point will be reached when further output growth will be held back. Figure 9.2 provides such a comparison between the rate of growth of Soviet industry and what may be termed the 'infrastructure' sectors. 'Infrastructure' includes: transportation, communications, construction, trade and distribution, and other services. (The data used here are Western estimates of GNP by sector of origin in constant prices.)[5]

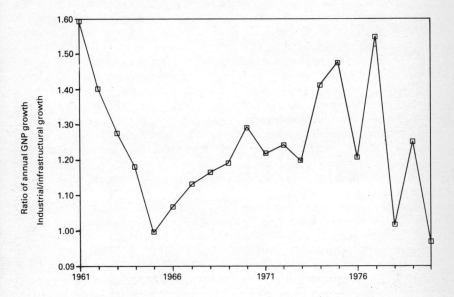

FIGURE 9.2   *Industrial and infrastructural growth*

A clear cycle is evident in figure 9.2. At the beginning of the 1960s, output growth in industry was considerably faster than the growth in infrastructure. Then there is a trend toward convergence between the two sectors, continuing until the mid-1960s. The two sets of growth rates then diverge once again, reaching a peak in the mid-1970s. After 1975 the pattern is erratic, with alternating annual swings up and down, but with a clear downward trend. Comparing the two cycles, it is striking how similar the movement of the growth rate ratios are in the first half of the 1960s and the second half of the 1970s. At the beginning of each period the rate of growth of industry was over 50 per cent greater than the rate of growth of infrastructure, and this was followed by a period when convergence between the two sectors was achieved. In the 1960s the growth of the two aggregates moved toward one another; industrial growth declined modestly while infrastructure growth grew. As a result, the aggregate growth rate remained high. In contrast, in the 1970s both industry and infrastructure had declining growth rates, and they converged at a low level of aggregate growth. The ability of Soviet planners to react in time to correct this situation was probably reduced by the overall decline in the rate of growth of investment that was decreed for the tenth Five-Year Plan period (1976–80).

The slow-down in Soviet output has continued into the early 1980s. In 1983 there was a rebound of gross domestic product (GDP) growth to an estimated 3 per cent, but this was followed by relatively poor performance in 1984. Thus the Soviet economy still seems to be struggling to escape from the malaise that appeared in the mid-1970s.

It should be pointed out that there was one area where the Soviets saw dramatic improvement in the 1970s: this was in foreign trade. Due primarily to sharp increases in world gold and oil prices, the Soviet Union reaped substantial 'wind-fall' gains from their exports to Western countries. These additional earnings helped them finance grain imports needed to make up for their harvest shortfalls and to increase greatly their imports of machinery and technology.[6] These machinery imports appear to have contributed only modestly so far towards increasing output, and they may in fact have been a factor contributing to the recent drop in growth rates in some sectors, as they caused temporary 'indigestion' in the system and were 'resource-demanding'. Using the figures on GDP growth given above, one might argue that there has been a negative correlation between these imports and increases in aggregate output. The problem one faces is in properly accounting for the time required for this imported capacity to be fully utilized.

A good example of this point is provided by the Soviet ferrous metals sector.[7] Output in this sector has persistently fallen below plan in recent years. Yet it is also a sector receiving a high volume of imported equipment. These machinery imports were focused on the finishing stage and thus could have no effect on the output of crude steel. But they could have improved the yield of finished steel per ton of crude steel. Yet, it appears that the problems of the sector may have been exacerbated in the short-run by the attempt to introduce substantial amounts of imported equipment, which drew away scarce material and human resources from existing plants. On top of this, much of the new capacity based upon these imports had still not been put into operation by the early 1980s. There could still be a turnaround in ferrous metals growth over the next few years as these plants are commissioned and the new technology begins to be assimilated.

RECENT SIGNS OF CHANGE

Results of the last few years show some improvement in the performance of the Soviet economy, particularly in key industrial sectors. At least the 'spiralling downward' of growth rates appears to have been arrested. But it is not clear whether the economy is settling into a 'low-growth plateau', or whether it is in the first stages of a more substantial growth recovery. Possible causes of this improvement are:

(1) Intersectoral growth has been more balanced as the performance of key bottleneck sectors, especially ferrous metals and railway transport, has improved.
(2) The campaign for worker and manager discipline initiated by Andropov appears to have resulted in some improvement in labour productivity, at least in the short-run.
(3) Investment is growing more rapidly than planned for the eleventh Five-Year Plan period. (Although a major change in investment policy has yet to be announced, there is a vigorous debate in progress inside the Soviet Union about investment policy, with many advocates for a higher level of investment.)

Some longer-term policy changes which have been initiated recently may provide eventual pay-offs:

(1) The Food Programme, begun under Brezhnev in 1982, shows promise of improving agro-industrial production, and could help smooth out some of the worst problems created by wide swings in agricultural output. This programme has realistically focused on improving the processing, storage and distribution of agricultural products rather than on increasing farm output.
(2) 'Tinkering' with the organization of planning and plan indicators continues, the latest round being the experiments begun in selected ministries in January of 1984. The latter is a small step in the direction of allowing more decentralized decision-making power for enterprise managers. There are signs that this type of change will be given wider scope in the twelfth Five-Year Plan period.
(3) Energy and material conservation show some signs of improvement, perhaps as a result of increases in fuel prices and the shift from gross to net output plan indicators that is being implemented as a result of the 1979 planning decrees. Although the results achieved to date are not dramatic, the possibilities for further improvement are substantial, as can be appreciated when comparisons are made with other industrialized economies.

While the Soviet Union's domestic economy has shown only modest improvement, its external economic position has remained strong. Although world energy prices have declined in the recent past, the Soviet Union has been able to increase the quantity of its oil exports so as to maintain growth in revenues. Arms sales have also contributed significantly to Soviet earnings (accounting for over 50 per cent of exports to developing countries), although much of this trade has required medium- and long-term loans to developing countries. Recent estimates show that the level of outstanding Soviet hard-currency loans to the Third World at the end of 1983 stood somewhere between $23 and $31 billion, most of which is linked to arms sales.[8] Even disregarding Soviet assets other than their deposits in Western banks, the level of Soviet net debt has declined over the past two years, and the Soviet debt service burden is low relative to export earnings.

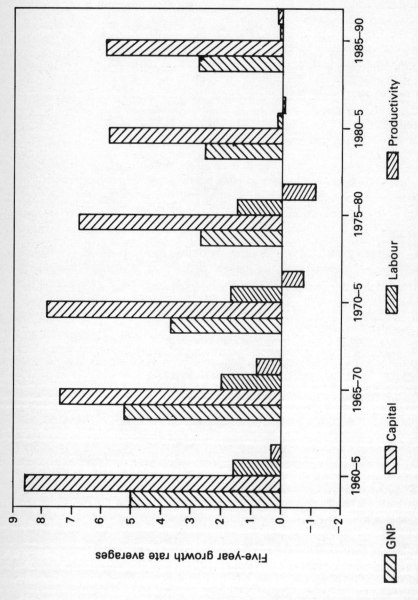

FIGURE 9.3 *Aggregate output, inputs and productivity*

THE MEDIUM-TERM OUTLOOK

There are generally two viewpoints concerning the post-1975 slow down in Soviet growth, and the prospects for future growth. On the one hand, the economic slow-down is seen as evidence of a dramatic shift downward in the long-term growth path of the economy, whilst on the other, the slow-down is seen as a temporary drop below the longer-term trend, which itself is one of gradually declining growth rates.[9] The forecasts presented below are consistent with the latter view.

If the 1976-82 growth slow-down was due primarily to bottlenecks in the economy, then the improvements that have recently been observed may be of a temporary nature. This would mean that further growth *acceleration* would not be expected. But if further bottlenecks can be avoided in the future, the Soviet economy should return to the growth path evident in the period up until 1976, rather than continue along the 1976-82 path of more rapidly declining growth rates. Based on this view of the economy, GDP can be expected to increase at the rate of 2.5 per cent to 3.0 per cent per year for the remainder of the decade. Given an average annual rate of increase in capital stock of almost 6 per cent and increases in labour supply of 0.2 per cent or less, this rate of GDP growth could be achieved with a very modest increase in aggregate factor productivity, as is depicted in figure 9.3. This would be an improvement over recent performance (aggregate factor productivity declined in the period 1975-80 by approximately 1 per cent per year) but below the results achieved in the 1960s.

The energy sector will continue to be one of the key factors in the outlook for both the domestic economy and foreign trade. There is a great deal of uncertainty in any forecast for Soviet energy production, consumption and trade, but tentative projections can be made. In the current Wharton forecast,[10] the 1983-4 period trends in production are expected to continue over the next few years. The most important of these has been the downturn in oil output from a peak in 1983, and the continued rapid growth of natural gas production. In addition, continued problems with coal are being encountered, as older capacity is being retired more rapidly than new capacity can come on-stream. This is coupled with the low quality and energy content of some of the newly tapped deposits and the inadequate infrastructure for their efficient exploitation. Further delays in the nuclear electric generation programme are also evident.

Oil production is projected to continue its decline, dropping at a rate of about 1 per cent per year. Coal production is also projected to experience steady decline through 1990, but output should at least level off during the 1990s as new capacity begins to keep pace with retirements. The natural gas picture provides the main growth potential for Soviet energy production. Even though output has grown at rates of 7-9 per cent over the last few years, current output is below capacity, limited by the rate at which the domestic economy can absorb the gas, and by the rate of potential exports to Eastern and Western Europe.

The cost advantage of natural gas production relative to oil production is considerable. According to available evidence and some very rough estimates, the marginal cost of producing and transporting Soviet natural gas is between one-fifth and one-quarter that for the energy equivalent in oil. Moreover, the marginal cost of gas production is believed to be close to the average cost, indicating that the cost advantage of gas over oil is growing rapidly. Coupled with the rather pessimistic outlook for oil prices, both on the world market and within the CMEA, it is logical that the Soviets should rethink their policy of expanding oil production at very high cost. They should opt instead for accelerated natural gas output and heightened effort to convert the domestic economy from the use of oil to natural gas. Even with the investment requirements of energy conversion, this strategy could be expected to produce better results in the long-term than a strategy in which there is continued effort to prevent the decline in the output of oil via very expensive recovery processes or a resort to new fields in increasingly inaccessible and inhospitable territory. Thus, in the baseline, Soviet natural gas production is projected to grow at over 7 per cent per year in the period 1986–90.

Another source of growth in Soviet energy production will be primary electric power. Within that sector, there have been bold plans for increasing the share of nuclear power generation in meeting the needs of the European USSR as the most cost-effective solution. However, despite considerable growth of nuclear generating capacity over the last 15 years, these plans have been grossly underfulfilled. At present, output of nuclear energy is equal to about 45–50 per cent of that of hydroelectric power. Nuclear generating capacity growth has been limited by construction delays and slow progress in completing the 'Atommash' reactor-building facility. We anticipate that these delays will postpone the surge in commissioned capacity until the early 1990s. Thus, in the forecast, production of primary energy is projected to grow at 5 per cent annually in the period 1986–90.

The assumed ratios of apparent energy consumption to net material product, used in the forecast, reflect a judgement that considerable progress can be made over this period in improving the energy efficiency of the economy. Between 1970 and 1980, despite interim developments in the world market price of energy, the energy/net material product ratio for the Soviet economy declined by only an average 0.9 per cent per year. Results in 1981–2 were little different. In 1983, however, a 1.9 per cent decline was observed. We project further decline in this ratio at 1.3 per cent per year through 1990. This is likely to be achieved on the basis of strictly administrative and planning measures (such as those recently used in East Germany) rather than through any sort of improved price and incentive mechanisms or significant structural changes in the economy.

The continued growth of energy production, albeit at somewhat reduced rates, combined with slower rates of growth of net material product and moderate improvements in the economy's efficiency in energy use, provides for continued growth in Soviet net energy exports. Total net exports in the forecast rise from 4.9 million barrels per day oil equivalent (b/doe) in 1984 to 6.3 million b/doe in 1990 (an average annual rate of growth of around 4 per cent).

The growth in net exports consists almost entirely of natural gas and some oil. The forecast for net energy exports by trading partner region reflects the Soviet strategy of maintaining growth of net oil exports for hard currency, supplemented by growing gas sales, with almost all of the increment in net exports to Eastern Europe taking the form of increased deliveries of gas.

In our view, these energy balance projections indicate that given the projected moderate growth in the Soviet economy, energy production would *not* baulk the growth of the domestic economy, nor require a dramatic change in Soviet trade relations either with Eastern Europe or the rest of the world.

Given the above conditions in the energy sector, the Soviet Union's external position is also likely to remain strong over the next few years, which would give the Soviets additional reserves in the event of unexpected problems in the domestic economy. Wharton's baseline forecast is for Soviet hard-currency, energy export earnings to climb from $23.6 billion in 1983 to $35 billion by 1990. Underlying this forecast is the projection of net Soviet oil exports to non-socialist countries, increasing from 1.6 million barrels per day in 1983 to 2.4 million barrels per day by 1988. (The world trade price of oil is projected to rise to $34 by 1990.) Gas exports to non-socialist countries are expected to rise even faster, from 22.6 billion cubic meters in 1983 to approximately 60 billion cubic meters in 1990.

The major question currently facing anyone attempting to forecast the level of Soviet imports from the West, is not what the Soviet Union can afford – their export earnings and access to Western credits are sufficient to allow them substantial lee-way in setting the level of imports. Nor is it difficult to project the composition of their imports. If a reasonable estimate can be made of grain imports, and this depends primarily on the harvest, then most of the remainder of Soviet imports will be industrial raw materials plus machinery and equipment. (It appears unlikely that the Soviet Union will change its policy of importing only limited amounts of manufactured consumer goods.) The real problem is in determining what the Soviet Union's policy will be toward relying on Western supplies of equipment and technology, and their views on the costs and benefits of these for their economy.

In recent years there has been some opposition in the Soviet Union towards increased imports from the West. Much of the equipment imported from the West has not produced the high returns expected, and there is concern about increased dependency at a time when the Western governments are actively using trade as an instrument of foreign policy. As Gary Bertsch points out in chapter 6, the Reagan Administration has made a concerted effort, with some success, to curtail all forms of technology transfer from the USA and its allies to the Soviet Union. This has led to Soviet concerns about the supply security of imported Western high-technology equipment and of the materials and services necessary to maintain its operation. However, as our forecast indicates, we feel that Soviet policy will more likely shift in the direction of greater trade with the West, rather than less. The results of last summer's CMEA summit and the recent actions and statements of Gorbachev seem to indicate that the Soviets will not attempt to curb their purchases in the West, but rather they are intent upon expanding them.

Given this policy perspective, and our forecasts of energy export revenues, we expect Soviet imports from non-socialist countries to be increasing at an average annual rate (1984–90) of almost 10.5 per cent in nominal terms (5.2 per cent in real terms), rising from $32.5 billion in 1984 to almost $60 billion in 1990. As we expect that imports of grain and other food products will decline, this means that imports of Western machinery and equipment could increase even faster. The average annual growth of machinery imports in the forecast is 19 per cent in nominal terms (13 per cent in real terms), with most of the increase coming in the early part of the next Five-Year Plan period.[11] It is our expectation that most of this growth in trade will be in imports from Western Europe, where there will be pressure on the Soviets to balance their trade due to large European purchases of Soviet gas. Japan will also be a major beneficiary, as the Soviets have increasingly turned to that country as an alternative source of high-technology equipment. Although these projections show a steadily rising level of external debt, the actual degree of debt burden (measured in terms of the level of debt servicing to exports) is declining.

There are, of course, numerous risks and uncertainties in these forecasts. Two important ones should be noted here. The major 'up-side risk' in this forecast is that it may underestimate the potential for increased productivity growth in the Soviet economy. Many Western analysts, including Amann and Dyker in this book, find little of promise in recent Soviet efforts to improve the functioning of their economy.[12] However, in a study of the 1979 economic decrees, Nancy Nimitz points out how the changes in planning and enterprise incentives that were introduced may provide the groundwork for improved performance by addressing some of the conditions necessary for increased innovation and technological advance in the producer-goods industries. But she also points out that the changes introduced stop short of forcing the economy towards greater technological change.[13]

It may be that the Soviet Union is now poised for further changes in what they call the 'economic mechanism'. The persistent sluggishness of Soviet growth since 1976 appears to have convinced Soviet leaders, including those in the military, that simply continuing with old methods and policies will not suffice. They also appear to recognize that the transition from an 'extensive growth' path to an 'intensive growth' path will be difficult to achieve without more fundamental changes than those experimented with in the past. Nimitz speculates that the problems facing the Soviet economy in the 1980s could 'be a blessing in disguise. If anything can motivate Soviet managers to accept the risks inherent in pursuing (rather than dodging away from) new technology, it is the recognition that the country is in real danger.'[14] However, David Dyker points out in chapter 8, there are many obstacles to the introduction of real reforms.[15] Moreover, it is unlikely that aggregate economic growth rates would improve quickly even if reforms were introduced. In fact, there would probably be a long period of relatively slow growth – perhaps even slower growth than at present – while the necessary adjustments were made. Even in a small economy such as that of Hungary, improvements in the economic mechanism still have not been reflected in improved aggregate growth rates several years after their introduction.

Another possible source of improved performance is greater utilization of imported capital and technology. As was pointed out above, the Soviet Union has been able to increase substantially its imports of machinery from the West over the last decade, and appears likely to be able to do so in the future. When combined with imports from Eastern Europe (which are increasing due to the deterioration in their terms of trade with the Soviet Union), the share of investment accounted for by imported machinery will probably continue to increase. Although there is little evidence so far that this imported equipment and technology has led to increased growth of the economy, there may still be a significant pay-off in the future.

While there is no way to predict the total impact of any of the above possibilities, an indication of their effect on overall growth is suggested by assuming that aggregate factor productivity improves at a rate comparable to that achieved in 1965–70. In this case, the rate of GDP growth would increase to 3.5 per cent per year on average for the second half of this decade.

The major 'down-side' risk would appear to relate to energy and trade. The baseline projection described above is optimistic in that it assumes continued high Soviet oil production, low growth in domestic and Eastern European oil requirements, and relatively rapid growth of demand for gas in Western Europe. Each of these has a fair chance of not happening. If hard-currency energy export-earning revenues decline, the Soviet Union could try to increase its non-fuel exports and/or increase its debt. But it is doubtful that the Soviets would be willing to increase their debt significantly, and they would also find it difficult to increase their non-fuel exports. Thus it is more likely that some reduction in imports (from the levels suggested by the above projections) would be required. Most of this reduction would probably be in imports of Western machinery and technology.

CONCLUSION

Soviet science and technology is undoubtedly making a contribution to the growth of the overall economy. But recent trends in the economy indicate that its contribution has not been adequate to overcome all of the negative trends otherwise retarding Soviet growth. Whether this is a temporary or longer-term problem is a matter of debate among specialists, as the various contributions to this book demonstrate. However, notwithstanding problems of motivation and departmental barriers to innovation, the Soviet economy is still sufficiently strong to allow any reasonable commitment of resources needed to support further advancements in science and technology. The Soviet Union has also been able to supplement its domestic technology with imported technology, although it is likely that it will not be as fortunate in this respect in the future as it has been in the past.

References

CHAPTER 1

1. R. Amann, J. M. Cooper and R. W. Davies (eds), *The Technological Level of Soviet Industry*, London and New Haven, Yale University Press, 1977, p. 66.
2. R. Amann and J. M. Cooper (eds), *Industrial Innovation in the Soviet Union*, London and New Haven, Yale University Press, 1982.
3. This interesting approach has been developed by S. Gomulka, *Inventive Activity, Diffusion and the Stages of Economic Growth*, Economics Institute, Aarhus University, Monograph 24, 1971. See R. Amann et al., *The Technological Level of Soviet Industry*, p. 22 for a brief criticism of Gomulka's approach.
4. Curiously, we are taken to task by Michael Boretsky for this very reason. See his review of our 1982 volume in *Journal of Comparative Economics*, Vol. 8, 1984, pp. 207–11.
5. In effect, the invitation to engage in fresh thinking about Soviet economic and political organization was extended by Andropov in a key article about Marx's teachings and socialist construction in the USSR in *Kommunist*, 1983, No. 3, pp. 9–23.
6. T. Zaslavskaya, 'The Novosibirsk report', *Survey*, Vol. 28, No. 1, Spring 1984, pp. 88–109.
7. R. W. Campbell, 'The Economy', in R. F. Byrnes (ed.), *After Brezhnev: Sources of Soviet Conduct in the 1980s*, London, Frances Pinter, 1983, p. 69.
8. P. Hanson, 'The plan fulfilment report for 1984', *Radio Liberty Research*, RL 37/85, 4 February 1985.
9. V. Trapeznikov, *Pravda*, 7 May 1982.
10. R. Leggett, 'Soviet investment policy in the 11th five year plan', in US Congress Joint Economic Committee, *Soviet Economy in the 1980s*, US Government Printing Office, Washington DC, 1983, p. 143. Kazimierz Poznanski advances two explanations of this phenomenon (1) that with a low level of per capita output, old installations remain a useful buffer; and (2) that Soviet workers prefer familiar technologies and are reluctant to retrain. See, K. Poznanski, 'The management of technological change in the Soviet Union and Eastern Europe', unpublished paper, June 1983, p. 71.
11. R. Leggett, ibid., table 2, p. 133.
12. R. W. Campbell, ibid., p. 70.
13. *Handbook of Economic Statistics*, CIA, 1984, p. 68.
14. Joint Economic Committee Briefing Paper, *USSR: Economic Trends and Policy Developments*, Office of Soviet Analysis, CIA, September 1983, tables 14 and 15, pp. 66–7.
15. R. Amann et al., *The Technological Level of Soviet Industry*, pp. 8–23.
16. A. Bergson, 'Technological Progress', in A. Bergson and H. S. Levine, *The Soviet Economy: Toward the Year 2000*, London, George Allen and Unwin, 1983, p. 65.
17. W. H. Cooper, 'Soviet–Western Trade', in US Congress Joint Economic Committee, *Soviet Economy in the 1980s*, p. 462; see also, E. A. Hewett, 'Foreign Economic Relations' in A. Bergson and H. S. Levine, *The Soviet Economy*, pp. 276–8. When allowance is made for energy prices and other price movements, the

rate of growth of machinery and equipment exports to the LDCs is above average, though their nominal share in total exports fell from 33.2 per cent in 1972 to 20.3 per cent in 1980: see T. A. Wolf, 'Changes in the pattern of Soviet trade with the CMEA and the non-socialist countries', in *External Economic Relations of CMEA Countries: their Significance and Impact in a Global Perspective*, NATO Economic Directorate, Brussels, 1983, pp. 222–5.

18 K. Poznanski, 'The management of technological change in the Soviet Union and Eastern Europe', unpublished paper, 1983, p. 19; 'Competition between Eastern Europe and the Developing Countries in the Western Market for Manufactured Goods', draft paper prepared by the same author for US Congress Joint Economic Committee, February, 1984.

19 V. Kontorovich, 'Technological progress and Soviet productivity slowdown', unpublished paper, March 1984. Kontorovich, a Soviet emigré with practical experience in the collection and processing of statistics, provides an interesting account of what these statistics *mean* and the relative biases to be found in them. Julian Cooper has made the point to me that *apparently* deteriorating performance could to some extent be accounted for by more objective statistical reporting – though this factor is hard to evaluate.

20 J. Cooper, 'Is there a technological gap between East and West?', paper presented to conference organized by the Canadian Institute of International Affairs, Toronto, June, 1984. To be published in J. Fedorowicz (ed.), *East–West Trade in the 1980s: Prospects and Policies*, Boulder, Colorado, Westview Press (provisional title).

21 J. A. Martens and J. P. Young, 'Soviet implementation of domestic inventions' in US Congress Joint Economic Committee, *Soviet Economy in a Time of Change*, Washington DC, US Government Printing Office, 1979, pp. 472–509.

22 As well as the growth in the manufacture of complex fertilizers there has been a marked growth in the average size of plants. See K. M. Dyumaev (Deputy Chairman, State Committee for Science and Technology), 'Vazhneishie zadachi razvitiya khimicheskoi nauki i promyshlennosti', *Khimicheskaya Promyshlennost'*, 1983, No. 2, pp. 3–5.

23 L. D. Bores, 'Agat: A Soviet Apple II Computer', *Byte*, November 1984, pp. 135–6 and 487–90.

24 K. Poznanski, 'The management of technological change in the Soviet Union and Eastern Europe', unpublished paper, June 1983, p. 20.

25 See, for example, an interesting draft paper by K. Poznanski, 'The extinguishing process: a case study of steel technologies in the world industry', November 1984.

26 J. Huxley, 'How ICI pulled itself into shape', *Sunday Times*, 29 July 1984, p. 57.

27 See, for example, K. M. Dyumaev, 'Vazhneishie zadachi', p. 5.

28 V. Yasmann, 'Personal computers in the Soviet Union', *Radio Liberty Research*, RL 308/84, 15 August 1984; P. Hanson, 'Soviet progress in the microprocessor field', *Radio Liberty Research*, 7 June 1984; J. Cooper, 'A note on microprocessors and microcomputers in the Soviet Union', internal CREES discussion paper, University of Birmingham, May 1982; V. Kuzmin and S. Klepikov, 'Electronics embargo: calculations and miscalculations', *Soviet News*, 29 August 1984; S. E. Goodman, *Technology Transfer and the Development of the Soviet Computer Industry*, draft report prepared for CSIS Task Force on Trade, Technology and Soviet–American Relations, Georgetown University, Washington DC, March 1984.

29 'Soviets drop further back in weapons technology', *Science*, Vol. 223, 16 March 1984 (quoting an assessment of Richard DeLauer, Chief Scientist, DOD). But, as a series of other assessments has made clear, a relative lag in technological sophistication *per se* does not necessarily imply a lag in operational effectiveness.

See, for example, J. W. Kehoe and K. S. Brower, 'US and Soviet Weapon System Design Practices', *International Defense Review*, Vol. 15, No. 6, 1982; statement by Major General Schuyler Bissell, Deputy Director of Defense Intelligence Agency in US Congress Joint Economic Committee, *Allocation of Resources in the Soviet Union and China – 1983*, Washington DC, US Government Printing Office, 1984, and statement of Henry Rowen, Chairman of National Intelligence Council, CIA in US Congress Joint Economic Committee, *Allocation of Resources in the Soviet Union and China –1982*, Washington DC, USGPO, 1983.

30 R. W. Campbell, *Soviet Technology Imports: the Gas Pipeline Case*, California Seminar on International Security and Foreign Policy, Discussion Paper 91, Santa Monica, California, February 1981; T. Gustafson, *The Soviet Gas Campaign: Politics and Policy in Soviet Decision-making*, Rand R-3036-AF, Santa Monica, June 1983.

31 J. Cooper, 'Industrial robots in the USSR', *CREES Discussion Paper*, University of Birmingham, 1980; J. Cooper, 'The application of industrial robots in the Soviet engineering industry, *Omega*, Vol. 12, No. 3, 1984, pp. 291–8.

32 J. Kiser, *Report on the Potential for Technology Transfer from the Soviet Union to the United States*, Office of External Research, Department of State, Washington DC, 1977; 'What gap? Which gap?' *Foreign Policy*, No. 32, Fall 1978, pp. 90–4; *Civilian and Military Technology in the USSR: What's the Difference?*, Kiser Research Inc., unpublished paper, May 1982 (examples include electroslag resmelting, electron beam guns, hard surface coatings to improve wear of turbine blades and tools, electro-impulse de-icing for aircraft; Kiser argues that most of these are spin-offs from the military sector). For another account of Soviet trade in licences see M. C. Spechler, 'Soviet policy towards technical change since 1975', US Congress Joint Economic Committee, *Soviet Economy in the 1980s*, pp. 89–101.

33 O. Bogomolov, 'Nauchno-tekhnicheskii progress v SSSR i ego vneshnepoliticheskie aspekty', *Planovoe khozyaistvo*, 1983, No. 4, p. 110.

34 Senior Researcher, Bundesinstitut für Ostwissenschaftliche und internationale Studien, Cologne, West Germany (point made during the proceedings of conference cited in f.n. 35).

35 R. Amann, 'Technical progress and political change in the Soviet Union' in *The CMEA Five-Year Plans (1981–1985) in a New Perspective*, NATO, Brussels, 1982, pp. 139–58; and 'The political and social implications of economic reform in the USSR', paper delivered at Second International Conference, *Interdependence of Economics and Politics in the USSR*, Cologne, West Germany, November 1984.

36 J. Cooper, 'Is there a technological gap', (June 1984).

37 B. P. Kurashvili, 'Sudby otraslevogo upravleniya', *Ekonomika i Organizatsiya Promyshlennogo Proizvodstva*, 1983, No. 10, pp. 34–55; G. Popov, 'The development of branch industrial management', *Kommunist*, 1982, No. 18, pp. 48–59 (abstracts in *Current Digest of the Soviet Press*, 6 July 1983).

38 V. Kudinov, *Pravda*, 21 July 1982. For a spirited reiteration of this point see G. Marchuk, 'Osnova intensifikatsii ekonomiki', *Partiinaya Zhizn'*, 1985, No. 1, p. 33.

39 G. Dobrov, *Pravda*, 8 June 1982.

40 V. Rassokhin, *Pravda*, 12 July 1982.

41 K. Puzynya, *Pravda*, 8 February 1983.

42 G. Marchuk, 'Nauchno-tekhnicheskii progress – osnova intensifikatsii obshchestvennogo proizvodstva', *Kommunist*, 1983, No. 4, p. 67.

43 S. Fortescue, *The Academy Re-organized: The R and D Role of the Soviet Academy of Science since 1961*, Occasional Paper No. 17, Department of Political Science, Research School of Social Sciences, Australian National University, Canberra, 1983, pp. 89–96; T. Gustafson, *Selling the Russians the Rope? Soviet Technology Policy*

and *US Export Controls*, Rand R-2649-ARPA, Santa Monica, April 1981. Chapter 6, pp. 51–65 is devoted to 'The increasing involvement of the Academy of Sciences in technological innovation'.
44 A. P. Aleksandrov, opening address to Annual General Meeting of the Soviet Academy of Sciences, *BBC Summary of World Broadcasts: the USSR*, 3 April 1984.
45 V. Kudinov, *Pravda*, 21 July 1982.
46 V. V. Listov, 'O merakh po uskoreniyu nauchno-tekhnicheskogo progressa v khimicheskoi promyshlennosti', *Khimicheskaya Promyshlennost'*, 1984, No. 4, pp. 4–5.
47 V. Fel'zenbaum, 'Upravlenie nauchno-tekhnicheskim progressom', *Voprosy Ekonomiki*, No. 11, 1983, p. 15.
48 V. Trapeznikov, *Pravda*, 7 May 1982.
49 T. M. Dzhafali, et al., 'Nekotorye aspekty uskoreniya nauchno-tekhnicheskogo progressa', *Sotsiologicheskie Issledovaniya*, 1983, No. 2, p. 59.
50 G. Marchuk, 'Nauchno-tekhnicheskii progress', p. 71.
51 L. E. Nolting and M. Feshbach, 'R and D employment in the USSR – definitions, statistics and comparisons', US Congress Joint Economic Committee, *Soviet Economy in a Time of Change*, pp. 746–7.
52 J. Cooper, 'Scientists and Soviet Industry: a Statistical Analysis', *CREES Discussion Paper*, Series RC/B17, November 1981, table 14, p. 45.
53 M. Korolev, *Pravda*, 25 March 1984.
54 See, for example, V. V. Listov, 'O merakh po uskoreniyu nauchno-tekhnicheskii progress', pp. 4 and 5. Ominously, CAD is identified as a task still 'standing before our designers'.
55 G. Lakhtin, *Pravda*, 26 November 1981.
56 A. Bergson, 'Technological progress', p. 64.
57 R. Leggett, 'Soviet investment policy', p. 133.
58 G. Marchuk, *Pravda*, 9 December 1983.
59 V. Fel'zenbaum, 'Upravlenie nauchno-tekhnicheskim progressom', p. 16.
60 V. K. Fal'tsman, 'Narodnokhozyaistvennyi zakaz na novuyu tekhniku', *Ekonomika i Organizatsiya Promyshlennogo Proizvodstva*, 1983, No. 7, pp. 3–19.
61 R. Kh. Vasil'eva, *Nesostoyatel'nost' burzhuaznykh traktovok ekonomicheskikh problem razvitogo sotsializma*, Kiev, 1983, p. 108.
62 Yu. Ya. Ol'sevich, *Ekonomicheskoe razvitie SSSR: Kritika burzhuaznykh kontseptsii*, Moscow, 1983, p. 197.
63 Yu. Ya. Ol'sevich, 'Kritika burzhuaznykh vzgladov na sotsialisticheskoe planirovanie nauchno-tekhnicheskogo progressa', *Planovoe Khozyaistvo*, 1984, No. 6, p. 106.
64 See, for example, T. M. Dzhafali, et al., 'Nekotorye aspekty uskoreniya nauchno-tekhnicheskogo progressa', *Sotsiologicheskie Issledovaniya*, 1983, No. 2, pp. 58–63 (the results of a survey of attitudes to reform on the part of managers, scientists and officials in Georgia).
65 M. S. Gorbachev, *Pravda*, 11 December 1984.
66 K. Chernenko, 'Na uroven' trebovanii razvitogo sotsializma: nekotorye aktual'nye problemy teorii, strategii i taktiki KPSS', *Kommunist*, 1984, No. 18, p. 9.

CHAPTER 2

1 *Materialy XXIV s"ezd KPSS*, Moscow, 1971, p. 46.
2 Dement'ev (aviation industry), *Izvestiya*, 22 May 1971; Butoma (ship-building),

Sel'skaya Zhizn', 26 June 1971; Bakhirev (machine-building), *Izvestiya*, 2 July 1971; Zverev (defence industry), *Izvestiya*, 7 July 1971.
3 A thorough search has failed to reveal a single work on the topic.
4 For a characteristically convoluted attempt to explain the 42 per cent see the statement of the US Defense Intelligence Agency in *Allocation of Resources in the Soviet Union and China – 1979*; Hearings before the Subcommittee on Priorities and Economy in Government of the Joint Economic Committee, US Congress, USGPO, Washington DC, 1980, p. 96.
5 *Pravda*, 30 October 1980.
6 S. Ventsov, *Narodnoe khozyaistvo i oborona SSSR*, Moscow, 1931, pp. 51–2; also S. Ventsov, *Za Industrializatsiyu*, 1 May 1930.
7 *Pravda*, 23 November 1946, article D. Ustinov, Minister for Armaments; *Pravda*, 1 June 1946 and 20 October 1946.
8 The Dnepropetrovsk truck factory was under construction from late 1945 with a planned capacity of 70,000 'ZIS' trucks per year (*Pravda*, 6 April 1946; *Fortune*, 1 August 1969, p. 122).
9 *Sotsialisticheskaya Industriya* (hereafter, *Sots. Ind.*), 15 July 1978. The first model was the 'MTZ-2' of the Minsk tractor factory. It is worth noting that the director of the Dnepropetrovsk 'Southern machine-building works' at that time was L. V. Smirnov, now Chairman of the Military-Industrial Commission (*Istoriya gorodov i sel Ukrainskoi SSR, Dnepropetrovskaya oblast'*, Kiev, 1977, p. 132).
10 *Pravda*, 9 August 1953, Malenkov's report to the USSR Supreme Soviet.
11 *Pravda*, 28 October 1953. This decree has been omitted from all the standard collections of government decrees published in recent years. See also the speech of Mikoyan, Minister of Trade, *Pravda*, 25 October 1953.
12 *Resheniya partii i pravitel'stva po khozyaistvennym voprosam*, Vol. 6, Moscow, 1968, pp. 563, and 158; Vol. 8, p. 430.
13 *Ekonomicheskaya Gazeta* (hereafter *Ekon. Gaz.*), 1970, No. 28, pp. 7–9.
14 *Materialy XXVI s"ezda KPSS*, Moscow, 1981, pp. 43–4.
15 *Ekonomicheskaya geografiya SSSR*, Moscow, 1976, p. 180. These wagons carry some 40 per cent of total Soviet rail freight (*Avtomatizirovannaya sistema upravleniya Uralvagonzavodom imeni F. E. Dzerzhinskogo*, Moscow, 1977, p. 5).
16 Trams are built by enterprises in Ust'-Katav, Riga and Leningrad (*Pravda*, 3 March 1983). The Riga and Leningrad works can be identified as belonging to civilian ministries. For MOM's involvement, see M. Agursky, *The Research Institute of Machine-Building Technology*, The Hebrew University of Jerusalem, The Soviet and East European Research Center, Soviet Institutions Series Paper No. 8, September 1976, p. 37; and *Pravda*, 20 October 1983.
17 *Pravda*, 24 August 1973; *Izvestiya*, 7 July 1971; *Pravda*, 7 March 1979; *Deputaty Verkhovnogo Soveta SSSR. Desyatyi sozyv*, Moscow, 1979, p. 190.
18 *Soviet Export*, 1979, 4(121), p. 38; *Ekon. Gaz.*, 1981, No. 7, p. 10.
19 *Izvestiya*, 7 July 1971; *Novye Tovary*, 1980, No. 10, p. 14; *Ekon. Gaz.*, 1978. No. 17, p. 17; *Novye Tovary*, 1979, No. 4, p. 10; *Kommercheskii Vestnik*, 1981, No. 1, p. 13; *Sovetskaya Rossiya*, 23 September 1984.
20 G. I. Medvedev, *Proizvodstvo tovarov narodnogo potrebleniya: Rezervy predpriyatii*, Leningrad, 1978, pp. 13–14; *Izvestiya*, 16 February 1980; *Pravda*, 1 December 1983.
21 From the total Soviet output of trucks and the known output of the Ministry of the Automobile Industry. In 1980 'other' production amounted to 43,000 units (*Narodnoe khozyaistvo SSSR, 1922–1982 gg*, Moscow, 1983, p. 196; and *Avtomobil'naya Promyshlennost'*, 1982, No. 8, p. 1).

22 *Sots. Ind.*, 1 October 1970; Yu. A. Lavrikov, E. V. Mazalov and A. P. Kuznetsov, *Ocherk ekonomicheskogo razvitiya Leningradskoi industrii za 1917–1967 gg.*, Leningrad, 1968, pp. 267–8; *Sots. Ind.*, 26 March 1985.
23 See ref. 9 above; *Sots. Ind.*, 15 November 1970.
24 The Ministry of Tractor and Agricultural Machine-building produces 85 per cent of all tractors (*Ekon. Gaz.*, 1982, No. 19, p. 2).
25 *Pravda*, 27 January 1981.
26 *Izvestiya*, 1 May 1982; *Pravda*, 3 December 1983; *Novye Tovary*, 1983, No. 10, p. 16; *Sovetskaya Torgovlya*, 11 September 1984.
27 *Sel'skaya Zhizn'*, 26 June 1971; *Sots. Ind.*, 17 July 1984.
28 *Ekon. Gaz.*, 1970, No. 28, p. 7; *Sots. Ind.*, 11 August 1972; *Sots. Ind.*, 1 October 1970.
29 *Ekon. Gaz.*, 1970, No. 28, p. 7; *Izvestiya*, 9 February 1984.
30 *Sel'skaya Zhizn'*, 26 June 1971; *Ekon. Gaz.*, 1982, No. 26, p. 2; *Pravda*, 1 July 1980; *Leningradskaya Pravda*, 11 August 1982; *Trud*, 20 July 1971 and 21 June 1983.
31 *Sots. Ind.*, 1 October 1970.
32 *Izvestiya*, 16 April 1978; *Izvestiya*, 4 January 1983; *Sots. Ind.*, 30 June 1984; *Leningradskaya Pravda*, 29 May 1979; Yu. A. Dmitriev, A. A. Zenkovich and R. F. Savinova, *Vladimir vchera, segodnya, zavtra*, Yaroslavl', 1981, p. 163; *Ekonomika i Organizatsiya Promyshlennogo Proizvodstva*, 1982, No. 1, p. 50; *Sudostroenie*, 1976, No. 5, p. 2; M. Agursky, 'The research institute of machine-building technology', pp. 62–71.
33 *Pravda*, 4 December 1979; *Mashinostroitel'*, 1981, No. 12, inside front cover.
34 *Izvestiya*, 12 February 1974; *Pravda*, 25 May 1976; *Ekonomika i Organizatsiya Promyshlennogo Proizvodstva*, 1982, No. 1, p. 72.
35 Julian Cooper, *Industrial Robots in the USSR*, CREES Discussion Paper, University of Birmingham, May 1980, pp. 25–7.
36 *Leningradskaya Pravda*, 17 December 1983; *Sots. Ind.*, 9 December 1982.
37 *Soviet Export*, 3(132), 1981, p. 17; *Pravda*, 25 May 1976, *Izvestiya*, 7 July 1971.
38 *Resheniya partii i pravitel'stva po khozyaistvennym voprosam*, Vol. 7, p. 522; *Sots. Ind.*, 15 October 1970; *Kommunist*, 1983, No. 4, p. 81; *Izvestiya*, 22 May 1971; *Sots. Ind.*, 10 January 1979; *Ekon. Gaz.*, 1977, No. 9, p. 6.
39 V. Tiunov, *Industrial'nye pyatiletki Zapadnogo Urala*, Perm', 1977, pp. 326, 384; *Sots. Ind.*, 11 August 1972; *Tekhnika i Nauka*, 1982, No. 3, p. 5.
40 *Izvestiya*, 17 July 1982; V. Tiunov, *Industrial'nye pyatiletki*, p. 383.
41 *Ekon. Gaz.*, 1971, No. 42, p. 6; *Leningradskaya Pravda*, 29 November 1981.
42 *Pravda*, 27 May 1983; *Izvestiya*, 14 April 1983.
43 *Sots. Ind.*, 7 July 1971; V. Tiunov, *Industrial'nye pyatiletki*, p. 383.
44 *Soviet Weekly*, 30 January 1982; *Pravda*, 3 August 1981.
45 *Ekon. Gaz.*, 1985, No. 2, p. 7; *Sots. Ind.*, 15 September 1983; *Pravda*, 10 October 1982; *Pravda*, 4 November 1981.
46 *Leningradskaya Pravda*, 6 June 1981.
47 *Izvestiya*, 2 July 1971; P. P. Karpov, *Raspredelenie sredstv proizvodstva v novykh usloviyakh khozyaistvovaniya*, Moscow, 1972, p. 62; *Razvitie i effektivnost' proizvodstva tovarov narodnogo potrebleniya*, Kiev, 1980, p. 273.
48 *Leningradskaya Pravda*, 6 June 1981 and 15 August 1984; *Sots. Ind.*, 21 June 1979; B. E. Tarasov and A. N. Emel'yanov, *Ekonomicheskii analiz nepreryvnoi razlivki stali*, Moscow, 1982, p. 6.
49 *Kuibyshevskaya oblast'*, Kuibyshev, 1977, p. 251; *Metallurgi (vchera i segodnya Kuibyshevskogo metallurgicheskogo zavoda im. V. I. Lenina)*, Kuibyshev, 1979.

50 H. Campbell, *Organization of Research, Development, and Production in the Soviet Computer Industry*, Rand, R-1617-PR, Santa Monica, December 1976, pp. 94 ff.; *Computing Surveys*, Vol. 10, No. 2, June 1978, p. 96.
51 *Mekhanizatsiya i Avtomatizatsiya Upravleniya*, 1983, No. 3, p. 7.
52 *Leningradskaya Pravda*, 15 November 1977; Kiev – deduced from known production of K580 microprocessor by Kiev 'Kristall'; *Pravda*, 27 December 1978; *Pravda*, 20 September 1981; *Sots. Ind.*, 11 January 1981.
53 *Izvestiya*, 7 July 1971; Ulyanovsk – author's deduction (see also M. L. Urban, *Soviet Land Power*, Ian Allan, London, 1985, p. 73).
54 Z. M. Balezin, *Shagi desyatiletii*, Kirov, 1981, p. 66.
55 *Sots. Ind.*, 17 November 1976; *Pravda*, 18 August 1977; *Kommercheskii Vestnik*, 1981, No. 24, p. 30; *Ekon. Gaz.*, 1982, No. 10, p. 23; *Izvestiya*, 29 July 1977; *Pravda*, 8 April 1981.
56 P. M. Stukolov, *Ekonomika elektronnoi promyshlennosti*, Moscow, 1976, p. 11.
57 *Pravda*, 20 September 1981; *Sots. Ind.*, 11 January 1981.
58 *Radio*, 1981, No. 7, pp. 1–3.
59 *Radio*, 1981, No. 1, pp. 2–3; *Ekon. Gaz.*, 1978, No. 25, p. 7.
60 *Soviet Export*, 1(148), 1984, p. 8; *Ekon. Gaz.*, 1982, No. 35, p. 10; V. P. Lomakin, *Primor'e vchera, segodnya, zavtra*, Moscow, 1981, p. 107.
61 *Resheniya partii i pravitel'stva po khozyaistvennym voprosam*, Vol. 6, Moscow, 1968, pp. 563–6.
62 *Sots. Ind.*, 16 April 1977; MOP as head ministry – no explicit reference, deduced by author.
63 *Literaturnaya Gazeta*, 9 September 1981, p. 10; *Sots. Ind.*, 18 September 1979; *Kommercheskii Vestnik*, 1983, No. 3, p. 4 (deduced from).
64 *Sots. Ind.*, 15 December 1982; M. Agursky, 'The research institute of machine building technology', p. 37; *Izvestiya*, 16 February 1980; *Novye Tovary*, 1976, No. 9, p. 4; *Sots. Ind.*, 16 April 1977; Yu. V. Subbotskii, 'Formy organizatsii industrii – novye yavleniya i problemy', *Znanie, Novoe v zhizni, nauki, tekhnike, seriya ekonomika i organizatsiya proizvodstva*, 1979, No. 6, p. 28; *Kommunist*, 1984, No. 10, p. 33; MM, MOP, MRP enterprises deduced from known affiliation of all other refrigerator-building enterprises.
65 *Izvestiya*, 16 February 1980; *Pravda*, 16 June 1981; V. P. Lomakin, *Primor'e vchera*, p. 104; *Sots. Ind.*, 14 October 1982; *Izvestiya*, 10 March 1985; *Novye Tovary*, 1976, No. 4, p. 25; *Izvestiya*, 2 March 1980; *Ekon. Gaz.*, 1983, No. 22, p. 8; *Izvestiya*, 10 August 1980; *Sots. Ind.*, 16 April 1977; *Izvestiya*, 4 October 1984.
66 *Kommercheskii Vestnik*, 1981, No. 18, p. 20; *Trud*, 20 November 1973; *Izvestiya*, 14 August 1980.
67 *Voprosy Ekonomiki*, 1983, No. 1, p. 103; *Pravda*, 10 January 1977; *Kommercheskii Vestnik*, 1981, No. 22, p. 28; *Izvestiya*, 10 June 1977.
68 *Sovetskaya Torgovlya*, 19 June 1984; *Pravda*, 4 January 1984; *Novye Tovary*, 1980, No. 6, p. 14.
69 *Pravda*, 24 October 1980; *Novye Tovary*, 1980, No. 7, p. 14; *Novye Tovary*, 1979, No. 3, p. 12; *Sots. Ind.*, 4 January 1978; *Trud*, 4 February 1978; *Pravda*, 5 July 1976.
70 *Sots. Ind.*, 6 April 1979; *Pravda*, 15 May 1979; *Izvestiya*, 16 February 1980; *Izvestiya*, 4 October 1977; *Trud*, 14 August 1979.
71 *Trud*, 3 January 1975; *Sots. Ind.*, 25 May 1977; *Kommercheskii Vestnik*, 1982, No. 23, pp. 24–5; *Novye Tovary*, 1977, No. 12, p. 4; *Kommercheskii Vestnik*, 1981, No. 9, pp. 12–13; *Novye Tovary*, 1980, No. 7, p. 23; *Izvestiya*, 16 February 1980; *Novye*

Tovary, 1978, No. 4, p. 20; *Kommercheskii Vestnik*, 1981, No. 2, p. 13 and 1983, No. 3, p. 4; *Novye Tovary*, 1979, No. 8, p. 5.

72 *Sots. Ind.*, 16 April 1977; *Pravda*, 18 October 1980; *Sovetskaya Torgovlya*, 31 July 1984; *Izvestiya*, 16 February 1980; *Standarty i Kachestva*, 1971, No. 6, p. 22.

73 *Sovetskaya Torgovlya*, 14 June 1984.

74 *Komsomol'skaya Pravda*, 4 June 1971; *Trud*, 18 December 1979; *Izvestiya*, 23 September 1981.

75 *Korabely – rodine*, Leningrad, 1981, pp. 104–5; *Trud*, 7 May 1971; M. Agursky, 'The research institute of machine-building technology', p. 37; *Novye Tovary*, 1975, No. 7, p. 9; *Izvestiya*, 9 February 1977; *Novye Tovary*, 1973, No. 9, p. 8; *Sots. Ind.*, 31 May 1979.

76 Yu. V. Subbotskii, 'Formy organizatsii industrii', p. 28.

77 US CIA, National Foreign Assessment Center, *Estimated Soviet Defense Spending: Trends and Prospects*, June 1970, USGPO, Washington DC, p. 1.

78 *Literaturnaya Gazeta*, 1981, No. 37 (9 September), p. 10. The concept of *grazhdanskoi produktsii* employed in the branch clearly excludes civilian aircraft.

79 *Izvestiya*, 16 February 1980; *Kommercheskii Vestnik*, 1982, No. 9, p. 9, and 1983, No. 3, p. 2. Note: the output of 'cultural and household goods' to which these data refer is frequently given in retail prices, i.e. inclusive of turnover tax. Sources rarely specify which prices are used, but here it is assumed that retail prices apply (explicit for MSP).

80 *Sots. Ind.*, 18 July 1971; *Izvestiya*, 16 February 1980.

81 *Kommercheskii Vestnik*, 1983, No. 22, p. 5; *Trud*, 1 January 1983; *Kommercheskii Vestnik*, 1983, No. 21, p. 2.

82 In 1977, 36 enterprises built refrigerators; in early 1984, only 18 (*Trud*, 9 September 1977; *Novye Tovary*, 1984, No. 7, pp. 26–7).

83 Deputy ministers include A. A. Kuz'mitskii (MPSS), V. Nikolaev (Minmash), S. S. Vinogradov (MSP) and E. A. Zhelonov (MOM), (*Izvestiya*, 18 July 1979; *Sots. Ind.*, 4 May 1983; *Kommercheskii Vestnik*, 1983, No. 3, p. 2; *Sovetskaya Torgovlya*, 19 June 1984).

84 *Kommercheskii Vestnik*, 1982, No. 9, p. 5.

85 *Pravda*, 29 September 1978; *Sovetskaya Torgovlya*, 13 January 1983 and 11 September 1984.

86 *Sovetskaya Torgovlya*, 13 January 1983.

87 *Ekon. Gaz.*, 1967, No. 49, p. 20; *Novye Tovary*, 1971, No. 12, p. 3; *Trud*, 2 August 1984; *Kommercheskii Vestnik*, 1983, No. 3, p. 3.

88 *Novye Tovary*, 1973, No. 3, p. 22.

89 According to Dunskaya, an emigré former employee of the Moscow Radio Factory (now the 'Temp' science-production association), most of the staff of the design sector concerned with 'Temp' television sets did not have security clearance; the sector was physically located outside the main premises of the factory design office (I. Dunskaya, *Security Practices at Soviet Scientific Research Facilities*, Delphic Associates Inc., Falls Church, Virginia, February 1983, p. 64).

90 See G. I. Medvedev, *Proizvodstvo tovarov*, pp. 52–3.

91 See, for example, D. E. Starik, F. I. Paramonov and I. I. Bugakov, *Ekonomika, organizatsiya i planirovaniya aviatsionnogo proizvodstva*, Moscow, 1976, p. 12.

92 V. S. Orlov, N. A. Mironova and I. N. Kolomitsev, *Formirovanie assortimenta elektrobytovykh tovarov*, Moscow, 1978, p. 46.

93 This raises a serious question. The creation of specialized facilities for the production of consumer and other civilian products at defence industry enterprises during recent years must have involved some expansion of enterprises and

the construction of new capacity. In assessing the rate of growth of the Soviet defence industry, the CIA, DIA and other agencies now place considerable emphasis on the growth of floor space at military production enterprises as measured from satellite reconnaissance photographs (see, e.g. US Department of Defense, *Soviet Military Power*, 3rd edn, April 1984, pp. 91–3). But do these agencies take account of the civilian activities of the enterprises and the creation of new floor space specifically for this purpose? If not, the claimed growth of floor space may overstate the real expansion of capacity for military production.

94 *Pravda*, 22 April 1978 and 24 August 1973.
95 *Sots. Ind.*, 18 July 1971.
96 *Leningradskaya Pravda*, 13 July 1983.
97 *Sots. Ind.*, 3 November 1970; *Pravda*, 7 March 1984.
98 *Ekonomika i Organizatsiya Promyshlennogo Proizvodstva*, 1984, No. 8, p. 88.
99 *Ekon. Gaz.*, 1970, No. 28, p. 7.
100 *Pravda*, 24 August 1973; *Ekon. Gaz.*, 1971, No. 42, p. 6; *Izvestiya*, 4 January 1983.
101 Supply problems experienced by the aviation industry (*Izvestiya*, 16 February 1980); the Leningrad 'Kirov factory' association (*Sovetskaya Rossiya*, 8 September 1984); the Dnepropetrovsk Brezhnev Southern machine-building works ('YuMZ' tractors) (*XXV s"ezd Kommunisticheskoi Partii Ukrainy. Stenograficheskii otchet*, Kiev, 1976, pp. 171–2 – speech of director, A. M. Makarov).
102 On the role of these military representatives, or *voenpredy*, see P. V. Sokolov (ed.), *Politicheskaya ekonomiya*, Moscow, 1974, p. 245.
103 See *Pravda*, editorial, 2 August 1984; *Pravda*, 12 December 1984; *Kommercheskii Vestnik*, 1984, No. 21, p. 19; *Pravda*, 27 January 1981.
104 *Pravda*, 5 February 1983; *Kommercheskii Vestnik*, 1981, No. 18, p. 20; *Ekon. Gaz.*, 1981, No. 39, p. 4; *Izvestiya*, 13 July 1983; *Kommercheskii Vestnik*, 1983, No. 23, p. 17.
105 *Sots. Ind.*, 18 July 1971; *Trud*, 3 December 1981; *Literaturnaya Gazeta*, 1981, No. 37, p. 10. Similarly, in the electronics industry in 1977 it was claimed that quality control systems used for basic production were being extended to consumer goods (*Radio*, 1977, No. 5, p. 8).
106 *Ekonomika i Organizatsiya Promyshlennogo Proizvodstva*, 1980, No. 16, p. 11; *Upravlenie kachestvom produktsii i effektivnost'yu resursov*, Moscow, 1980, p. 19; *Komsomol'skaya Pravda*, 4 June 1971.
107 *Ekon. Gaz.*, 1979, No. 24, p. 1.
108 *Sots. Ind.*, 12 December 1979; *Trud*, 16 June 1982; *Kommercheskii Vestnik*, 1984, No. 4, p. 22; *Trud*, 14 August 1979; *Pravda*, 27 February 1982.
109 *Kommercheskii Vestnik*, 1983, No. 3, p. 3.
110 *Planovoe Khozyaistvo*, 1978, No. 10, p. 23; *Trud*, 18 December 1979; *Kommunist*, 1984, No. 10, p. 34.
111 *Pravda*, 20 January 1985; *Ekon. Gaz.*, 1984, No. 30, p. 18.
112 See, for example, *Trud*, 1 April 1983; *Sovetskaya Rossiya*, 8 February 1984.
113 *Pravda*, 15 September 1984; *Leningradskaya Pravda*, 17 November 1984.
114 *Sots. Ind.*, 28 August 1971; *Pravda*, 4 January 1984; *Sots. Ind.*, 25 July 1984; *Pravda*, 12 February 1985, 19 April 1985 and 20 May 1985. The models concerned are the 'Berezka' produced by the Khar'kov 'Kommunar' association, and the 'Slavutich' produced by the 'Kiev radio factory' association.
115 *Kommercheskii Vestnik*, 1982, No. 16, p. 16; *Soviet Export*, 2(149), 1984, p. 18; 6(123), 1979, p. 19.
116 *Soviet Export*, 6(135), 1981, p. 8; *Soviet Export, Statistics*, 4 (supplement), 1981, p. 18; *Soviet Export*, 4(133), 1981, p. 47; 1(148), 1984, p. 10; 3(132), 1981, p. 19.

117 *Ekon. Gaz.*, 1976, No. 26, p. 8.
118 I. Dunskaya, 'Security practices', pp. 15, 63–5.
119 *Sovetskaya Rossiya*, 5 April 1972.
120 Thus the Nizhnii-Tagil 'Uralvagonzavod' produces approximately 20,000 freight wagons per year, 'Izhmash' almost 400,000 motorcycles, the Krasnoyarsk factory up to 750,000 refrigerators (1985 plan), the Penza 'Frunze factory' association one million bicycles, and the Khar'kov electrical apparatus factory (MOM?) 2.5 million electric razors (*Ekon. Gaz.*, 1984, No. 22, p. 8; *BBC SWB* SU/W1327/A/10, 22 February 1985; *Kommercheskii Vestnik*, 1982, No. 16, p. 16; 1984, No. 16, p. 20 and 1984, No. 23, p. 12).
121 *Sots. Ind.*, 23 May 1984; *Pravda*, 20 April 1985.
122 *Ekon. Gaz.*, 1970, No. 28, p. 7 (Brezhnev).
123 See, for example, the case of Minobshchemash and the production of gas stoves (*Pravda*, 10 January 1977).
124 *Sots. Ind.*, 9 March 1981; *Kommunist*, 1984, No. 10, p. 35 (Kapitonov, Central Committee Secretary with responsibility for consumer goods production).
125 One of the few works to examine specifically the issue of transfers is R. W. Campbell, 'Management spillovers from Soviet space and military programmes', *Soviet Studies*, Vol. 23, No. 4, April 1972, pp. 586–607. Campbell concluded that certain managerial practices can be transferred, but expressed scepticism about the possibility of technological spin-offs.
126 *Sots. Ind.*, 19 January 1985; *Ekon. Gaz.*, 1984, No. 23, p. 24.
127 *Sots. Ind.*, 2 October 1982.
128 *Pravda*, 14 February 1985; *Leningradskaya Pravda*, 15 March 1985.
129 See S. Melman, *Barriers to Conversion from Military to Civilian Industry – in Market, Planned and Developing Economies*, Paper prepared for the United Nations Centre for Disarmament, April 1980.

CHAPTER 3

1 L. A. Leventhal, *Introduction to Microprocessors: Software, Hardware, Programming*, USA, Prentice-Hall, 1978, pp. 17–18.
2 F. Faggin and M. E. Hoff, 'Standard parts and custom design merge in four chip processor kit', in L. Altman (ed.), *Microprocessors (Electronics Books Series)*, New York, McGraw Hill, 1975, p. 2.
3 J. Northcott, J. Martix and A. Zeilinger, *Microprocessors in Manufacturing Products*, Policy Studies Institute (PSI), November 1980.
4 J. Bessant, *Microprocessors in Production Processes*, PSI, No. 609, July 1982.
5 Ibid.
6 See, for example, A. Romanov, 'Mikro-EVM prikhodit v tsekh', *Izvestiya*, 14 October 1983.
7 J. S. Berliner, *The Innovation Decision in Soviet Industry*, Cambridge, Mass, 1976; and R. Amann and J. M. Cooper (eds), *Industrial Innovation in the Soviet Union*, London and New Haven, Yale University Press, 1982.
8 See P. Large, 'Computer men lured by Russia', *Guardian*, 3 February 1984.
9 J. Northcott and P. Rogers, *Microelectronics in Industry: What's happening in Britain*, PSI, No. 603, March 1982.
10 See R. Wohl, 'Soviet research and development', *Defense Science and Electronics*, September 1983, p. 11.

11 See, 'Why Russia is reluctant to boot up', *The Economist*, 5 May 1984, p. 58.
12 See, R. Heuertz, 'Soviet Microprocessors and Microcomputers', *Byte*, April 1984, pp. 351–62.
13 L. A. Zalmanzon, E. I. Pupyrev and V. I. Prangishvili, 'Mikroprotsessory v upravlenii i svyazi' *Znanie. Novoe v zhizni, nauke, tekhnike: seriya 'Radioelektronika i svyazi'*, No. 7, 1982.
14 See, *Radio Fernsehen Elektronik*, Vol. 31, No. 5, May 1982, pp. 280–3.
15 A. V. Giglavyi et al., *Mikro-EVM SM 1800: Arkhitektura, Programmirovanie, Primenenie*, Moscow, 1984, pp. 132–4.
16 E. P. Velikhov, 'Personal'nye EVM – segodnyashnyaya praktika i perspektivy', *Vestnik Akademii Nauk SSSR*, 1984, No. 8, p. 5.
17 A. A. Vasenkov, 'Razvitie mikroprotsessorov i mikro-EVM semeistva "Elektronika-NTs" na osnove kompleksno-tselevykh programm', *Elektronnaya Promyshlennost'*, November 1979, p. 16.
18 Ibid., p. 14. This refers to the NTs-80; M. F. Gal'perin and V. V. Gorodetskii, *Upravlyayushchie Sistemy i Mashiny*, 1982, No. 6, pp. 17–24. This identifies the NTs-80 as the K1801VE1 microcomputer.
19 See, for example, 'Annual update of microprocessors', *EDN*, 5 November 1980, p. 180.
20 Soviet author's certificate number 746532. Declared 24 April 1978, Published 7 July 1980, Authors: E. P. Balashov, G. Ya. Kuz'min, M. S. Kupriyanov and D. V. Puzankov.
21 E. P. Balashov and D. V. Puzankov, *Mikroprotsessory i mikroprotsessornye sistemy*, Moscow, 1982, p. 292.
22 D. Kennett, 'East German firm plans to export microcomputer system', *Mini-Micro Systems*, June 1982, pp. 120–3.
23 A. A. Vasenkov and V. A. Shakhnov, *Mikroprotsessornye komplekty integral'nykh skhem*, Moscow, 1982, pp. 163–73.
24 Ibid.
25 *Radio Fernsehen Elektronik*, Vol. 31, No. 5, May 1982, pp. 280–3.
26 S. Khristova et al., 'Problemno-orientirovannyi mikrokomp'yuter 1ZOT 100Zs' in *Proceedings of Symposium on the Application of Microprocessors and Microcomputers*, October 1979, Budapest, pp. 391–8.
27 *Tekhnika i Nauka*, 1984, No. 12, p. 11.
28 See below, p. 64.
29 'Why Russia is reluctant to boot up', *The Economist*, 5 May 1984, p. 58.
30 A. A. Vasenkov, 'Razvitie mikroprotsessorov', p. 17.
31 E. P. Balashov and D. V. Puzankov, *Mikroprotsessory*, pp. 307–17.
32 A. A. Vasenkov and V. A. Shakhnov, *Mikroprotsessornye komplekty*, pp. 70–80 and 179–83.
33 R. Heuertz, 'Soviet microprocessors', p. 358.
34 Yu. E. Golyas and V. A. Tikhonov, 'Mikroprotsessornye ustroistva v radiotekhnike', *Radiotekhnika*, 1983, No. 1, p. 11.
35 Ibid.
36 A. A. Vasenkov, 'Razvitie mikroprotsessorov', pp. 13–17.
37 M. P. Gal'perin and V. V. Gorodetskii, *Upravlyayushchie Sistemy, i Mashiny* , 1982, No. 5, pp. 105–8.
38 A. V. Giglavyi et al., *Mikro-EVM*, p. 5. This whole book is concerned with the SM 1800 microcomputer.
39 Ibid., pp. 132–4.
40 For example, see D. Kennett, 'East German firm plans'.

41 N. B. Mozhaeva, 'Mikroprotsessory v narodnom khozyaistve', *Pribory i Sistemy Upravleniya*, 1983, No. 12, p. 38.
42 E. P. Velikhov, 'Personal'nye EVM'.
43 Ibid.; L. D. Bores, 'Agat a Soviet Apple II computer', *Byte*, November 1984, pp. 135–6, 486–90.
44 R. Svoren', 'Nuzhen li personal'no vam personal'nyi komp'yuter?' *Nauka i Zhizn'*, 1984, No. 10, p. 69.
45 *Moskovskaya Pravda*, 12 May 1984.
46 E. P. Velikhov, 'Personal'nye EVM', pp. 6–8.
47 Ibid., p. 6.
48 *XXIV s"ezd Kommunisticheskoi Partii Sovetskogo Soyuza, Stenograficheskii Otchet*, Moscow, Vol. 1, 1981, pp. 272–6.
49 L. A. Zalmanzon, E. I. Pupyrev and V. I. Prangishvili, 'Mikroprotsessory v upravlenii', p. 4.
50 *Ekonomicheskaya Gazeta*, 1981, No. 5, p. 11.
51 Ibid.
52 P. M. Stukolov, Letter in *Ekonomicheskaya Gazeta*, 1982, No. 24, p. 2.
53 A. N. Chebotarev, 'Avtomatizirovannaya sistema upravleniya skladskimi protsessami na baze mikro-EVM', in *Mekhanizatsiya i Avtomatizatsiya Proizvodstva*, 1983, No. 3, pp. 8–10.
54 This is from a business source and is unattributable.
55 B. Bal'mont, 'Istoki obnovleniya. Marshruty tekhnicheskogo progressa', *Pravda*, 17 May 1984.
56 *Ekonomika i Organizatsiya Promyshlennogo Proizvodstva*, 1982, No. 1, pp. 47–87.
57 P. Hanson, 'Soviet progress in the microcomputer field', *Radio Liberty Research Bulletin*, RL 228/84, 7 June 1984.
58 J. G. Posa, 'Soviet chips feature refined fabrication but mimic US ICs', *Electronics*, 27 January 1981, pp. 39–40.
59 Ibid.
60 'Russian chip in US hands', *Electronics Weekly*, 19 November 1980, p. 11; J. Stansell, 'Russian chip-makers catch up', *New Scientist*, 27 November 1980, p. 557.
61 Including, *Pribory i Sistemy Upravleniya, Mekhanizatsiya i Avtomatizatsiya Proizvodstva, Radiotekhnika* and *Mekhanizatsiya i Avtomatizatsiya Upravleniya*.
62 V. M. Proleiko, 'Razvitie mikroprotsessorov, mikro-EVM sistem na ikh osnove', *Elektronnaya Promyshlennost'*, November 1979, pp. 3–6.
63 V. Ya. Kuznetsov, 'Odnoplatnaya mikro-EVM 'Elektronika-S5-21'' in *Proceedings of the symposium on microcomputer and microprocessor application*, Budapest, 17–19 October 1979, Vol. 1, OMKDK Techninform, pp. 41–7.
64 D. Zhimerin, *Izvestiya*, 12 August 1982.
65 M. Rakovskii, 'Novoi etap v razvitii mikroelektronika dlya sredstv vychislitel'noi tekhniki', *Ekonomicheskoe Sotrudnichestvo Stran-Chlenov SEV*, 1982, No. 9, pp. 44–6.
66 See, for example, 'CoCom rules . . . OK?', *Financial Times. East European Markets*, 23 July 1984, pp. 1–3.
67 See D. Kennett, 'East German firm plans'.
68 *Pravda*, 13 March 1984.
69 S. Bogatko and N. Morozov, 'Kamaz: Put' k sovershenstvu. 2: O chustve tseli', *Pravda*, 10 July 1984.
70 See E. P. Velikhov, 'Personal'nye EVM'.
71 See V. Yasmann, 'Personal Computers in the Soviet Union: Technology and Politics', *Radio Liberty Research Bulletin*, RL 308/84, 14 August 1984; and L.

Graham, 'Computers challenge the Soviet System', *International Herald Tribune*, 5 April 1984; and 'Circuits vs Soviet hard lines', ibid., 6 April 1984.
72 The ministries said to be involved in the production of PCs in the USSR are *Minradprom* and *Minpribor*, (E. P. Velikhov).
73 *Moskovskaya Pravda*, 12 May 1984.
74 'E/W – USSR linked with foreign electronic data processing systems', from an *A-Wire* report, *CND74*, 29 March 1984, based on a radio report from Radio Moscow on 28 March.
75 M. May, 'Russia in talks on ICL factory that could beat technology embargo', *The Times*, 4 February 1985.
76 A. A. Vasenkov and V. A. Shakhnov, *Mikroprotsessornye komplekty*, p. 178.
77 'Russian chip in US hands', *Electronics Weekly*, 19 November 1980, p. 11.
78 A. A. Vasenkov and V. A. Shakhnov, *Mikroprotsessornye komplekty*.
79 Ibid., pp. 178–87.
80 Ibid.
81 'Annual update of microprocessors', *EDN*, 5 November 1980, p. 102.
82 'World sales of microprocessors', *Electronics*, 13 January 1982, p. 124.
83 Ibid., p. 126.
84 M. Ravkovskii, 'Novoi etap'.
85 Ibid., p. 45.
86 Ibid.

CHAPTER 4

1 In 1978 it was decided to invest 2.9 billion rubles ($4.2 billion at official 1978 exchange rates) in the microbiological industry over the period 1981–5 – *Ekonomicheskaya Gazeta*, 1978, No. 30, p. 4.
2 A. H. Rose, 'History and scientific basis of large-scale production of microbial biomass', in A. H. Rose (ed.), *Microbial Biomass*, Academic Press, 1979.
3 See L. A. Cole, 'Yellow rain or yellow journalism?', *Bulletin of the Atomic Scientists*, Vol. 40, No. 7, August–September 1985.
4 K. Done, 'Bugs in the challenge to soyabeans and fishmeal', *Financial Times*, 20 July 1978.
5 A. I. Kozlov and Ya. V. Epshtein, 'Gidroliznaya promyshlennost' k 60-letiyu Velikogo Oktyabrya', *Gidroliznoe Proizvodstvo. Sbornik Referativnykh Materialov (GPSRM)*, 11(100), 1977, p. 3: A. A. Pokrovskii, *Belki odnokletochnykh organizmov mediko – biologicheskaya otsenka i perspektivy ispol'zovaniya v narodnom khozyaistve*, Moscow, 1971, p. 9; A. A. Pokrovskii, 'Perspektivy ispol'zovaniya belkov odnokletochnykh organizmov' in A. A. Pokrovskii (ed.), *Mediko-biologicheskie issledovaniya uglevodorodnykh drozhzhei*, Moscow, 1972, p. 13.
6 A. I. Kozlov and Ya. V. Epshtein, 'Gidroliznaya promyshlennost'.
7 Ibid.; A. A. Pokrovskii (ed.), *Mediko-biologicheskie issledovaniya*.
8 A. I. Kozlov and Ya. V. Epshtein, 'Gidroliznaya promyshlennost'.
9 A. H. Rose, 'The microbial production of food and drink', *Scientific American*, Vol. 245, No. 3, 1981.
10 A. H. Rose (ed.), *Microbial biomass*.
11 A. I. Kozlov and Ya. V. Epshtein, 'Gidroliznaya promyshlennost'.
12 Ibid.
13 A. E. Humphrey, 'Single cell protein: a case study in the utilization of technology in the Soviet system', in J. R. Thomas and U. M. Kruse-Vaucienne (eds), *Soviet*

Science and Technology: Domestic and Foreign Perspectives, Washington, DC, George Washington University, 1977.
14 A. I. Kozlov and Ya. V. Epshtein, 'Gidroliznaya promyshlennost'.
15 R. V. Katrush and E. G. Mirzayanova, 'Industriya belka', Zhurnal Vsesoyuznogo Khimicheskogo Obshchestva im. D. I. Mendeleeva, Vol. 27, No. 6, 1982, p. 18.
16 A. E. Humphrey, 'Single cell protein'.
17 Ibid.
18 Ibid.
19 Ibid., and R. V. Katrush and E. G. Mirzayanova, 'Industriya belka'.
20 S. P. Kovalenko, 'Problema kormovogo i pishchevogo belka mikroorganizmov', in Mikroorganizmy v promyshlennosti i sel'skom khozyaistve, Minsk, 1975, p. 29.
21 A. Pokrovsky, 'Some results of SCP medico-biological investigations', in S. R. Tannenbaum and D. I. C. Wange (eds), Single Cell Protein II, MIT Press, Cambridge, Mass, 1975.
22 R. V. Katrush and E. G. Mirzayanova, 'Industriya belka'.
23 Moscow Narodny Bank Press Bulletin, 17 September 1980, p. 6.
24 A. E. Humphrey, 'Single cell protein'.
25 R. Rychkov, 'What microbiology can do – young branch makes strides', Pravda, 2 June 1980, translated in Current Digest of the Soviet Press (CDSP), Vol. XXXII, No. 22, 1980.
26 L. N. Korshunova and N. I. Vakherkina, 'Opyt raboty s kadrami na Novopolotskom zavode BVK', in Mikrobiologicheskaya Promyshlennost' – Nauchno-tekhnicheskii Referativnyi Sbornik (MPNTRS), 2(188), 1983, p. 24; BBC Summary of World Broadcasts (SWB), SU/W1123/A/9, 27 February 1981.
27 Y. Skomorokhin, 'Sostavnoe zveno agropromyshlennogo kompleksa' in Ekonomicheskoe Sotrudnichestvo Stran-Chlenov SEV, 1984, No. 1, p. 38.
28 Ibid.
29 P. A. Bobrovskii, V. I. Kuz'min and G. M. Kaufman, 'Vliyanie syr'evykh zatrat na sebestoimost' belkogo-vitaminyakh kontsentratov (BVK)' in MPNTRS, 2(133), 1976, p. 32.
30 V. P. Aristova, L. P. Khayurova and A. V. Skryabina, 'O gigroskopicheskikh svoistvakh sutkikh kormovykh drozhzhei polyuchennykh na osnove ochishchennykh zhidkikh parafinov nefti', Mikrobiologicheskii Sintez, No. 5, 1967, p. 5.
31 P. A. Bobrovskii et al., 'Vliyanie syr'evykh zatrat'.
32 A. Einsele, 'Biomass from higher n-alkanes' in H. Dellweg (ed.), Biotechnology, Vol. 3, Weinheim, Basel, Verlag Chemie, 1983, p. 47.
33 A. E. Humphrey, 'Single cell protein'.
34 W. Dimmling and R. Seipenbusch, 'Raw materials for production of SCP', Process Biochemistry, Vol. 13, No. 3, 1978.
35 L. G. Sapozhnikov, 'Kirishi biokhimicheskii zavod' in MPNTRS, 7(149), 1977.
36 A. Einsele, 'Biomass from higher n-alkanes'.
37 Sotsialisticheskaya Industriya, 25 July 1984.
38 Sotsialisticheskaya Industriya, 25 July 1984.
39 SWB: SU/W1123/A/9, 27 February 1981.
40 P. A. Bobrovskii et al., 'Vliyanie syr'evykh zatrat'.
41 Ibid., pp. 33–4.
42 L. G. Sapozhnikov, 'Kirishi biokhimicheskii zavod'.
43 N. V. Ivanova, 'Etapy osvoeniya' in MPNTRS, 6(180), 1981, p. 29.
44 Ibid.

45 R. Rychkov, 'Mikrobiologicheskaya promyshlennost' v sisteme APK', *Ekonomika Sel'skogo Khozyaistva*, 1984, No. 4, pp. 55–61.
46 T. N. Kazakova, 'The granulation of feed yeast', *Hydrolysis and Wood Chemistry*, No. 2, 1983, pp. 25–6 (cover-to-cover translation of *Gidroliznaya i Lesokhimicheskaya Promyshlennost'*).
47 *Sotsialisticheskaya Industriya*, 25 July 1984.
48 Ibid., Soviet state standards are obligatory for Soviet producers, not advisory.
49 Ibid.
50 R. Rychkov, 'Mikrobiologicheskaya promyshlennost'.
51 G. M. Mukhaemetova, A. A. Kurmayeva, R. M. Khayrullina, T. S. Alibayev and F. A. Yarmukhaemetova, 'Some problems of labour hygiene and the state of health of workers producing protein-vitamin concentrates', *Gigiyena Truda i Professional'nye Zabolevania*, No. 9, 1979, pp. 23–5 (translated by *Joint Publications Research Service*).
52 Yu. V. Koval'ski, N. B. Gradova, V. R. Arkhipova and S. A. Konovalov, 'Microbiologicheskaya kharakteristika gotovogo produkta sredakh s n-parafinami' in *MPNTRS*, 7(127), 1975, p. 12.
53 Ibid.
54 Ibid.
55 Ibid., pp. 12–13.
56 Ibid., p. 13.
57 Ibid.
58 Ibid., p. 12.
59 A. Rimmington, 'Biotechnology in the USSR', *Industrial Biotechnology Wales (Biotechnolog Diwydiannol Cymru)*, Vol. 3, No. 8, 1984, pp. 1–5.
60 N. V. Ivanova, 'Etapy osvoeniya'.
61 Ibid.
62 N. B. Dem'yanovich, 'Organizatsiya sotsialisticheskogo sorevnovaniya v tsekhe tsentral'noi laboratorii izmeritel'noi tekhniki Krasnodarskogo khimicheskogo kombinata' in *MPNTRS*, 3(173), 1980, p. 30.
63 *Sotsialisticheskaya Industriya*, 25 July 1984.
64 Ibid.
65 *Ekonomicheskaya Gazeta*, 1983, No. 51, p. 2.
66 Ibid.
67 R. Rychkov, 'Mikrobiologicheskaya promyshlennost'.
68 Ibid.
69 *Ekonomicheskaya Gazeta*, 1983, No. 51, p. 2.
70 G. B. Carter, 'Is biotechnology feeding the Russians?', *New Scientist*, 23 April 1981.
71 R. L. Sopko (ed.), *Pulp and Paper in the USSR*, Dirosab, Farsta, Sweden, 1976, p. 43.
72 V. I. Kropotov, 'Expanding the production from the plants of the pulp and paper industry', *Hydrolysis and Wood Chemistry*, 1978, No. 6, pp. 1–3.
73 R. L. Sopko, *Pulp and paper in the USSR*.
74 V. Faust and P. Präve, 'Biomass from methane and methanol', in H. Dellweg (ed.), *Biotechnology*, Vol. 3, p. 104.
75 V. I. Kropotov, 'Expanding the production'.
76 M. Ringfeil, 'SCP from carbohydrate-containing effluents and animal production wastes – the state of the art in the GDR' in *International Symposium on Single Cell Proteins* (Paris; January 28, 29, 30, 1981), Technique et Documentation, Lavoisier, 1983, p. 274.

77 H. Rothman, *Biotechnology: a Synoptic Review*, London, 1980, p. 17.
78 A. F. Demin, 'Economy of feedstock and materials', *Hydrolysis and Wood Chemistry*, 1981, No. 1, pp. 21–2.
79 T. M. Semushina, 'Maximum attention to the selection and introduction of high-yielding strains', *Hydrolysis and Wood Chemistry*, 1977, No. 1, pp. 3–4.
80 Ibid.
81 V. I. Kamennyi, 'The path of efficiency and quality', *Hydrolysis and Wood Chemistry*, 1981, No. 5, pp. 1–4.
82 T. N. Semushina, 'Maximum attention'.
83 E. L. Golovlev, 'Study of solid-phase fermentation of plant materials in the USSR' in *International Symposium on Single Cell Proteins*, p. 245.
84 V. I. Kamennyi, 'The path of efficiency and quality'.
85 A. E. Humphrey, 'Single cell protein'.
86 P. Hanson, 'The Soviet chemical industry's response to agriculture's needs' *Outlook on Agriculture*, Vol. 12, No. 4, 1983, p. 200.
87 *BBC Summary of World Broadcasts* (*SWB*): SU/W1270/A/11, 13 January 1984; *SWB*: SU/W1273/A/14, 3 February 1984; *SWB*: SU/W1268/A/8, 23 December 1983; *SWB*: SU/W1275/A/9, 17 February 1984; R. Rychkov, 'Mikrobiologicheskaya promyshlennost'.
88 Ibid.
89 Ibid.
90 M. Vasin, *Pravda*, 3 January 1982, translated in *CDSP*, Vol. XXXIV, No. 1, 1982.
91 C. Cookson, *The Times*, 1 February 1983; *East–West (Fortnightly Bulletin)*, 28 April 1983, No. 314, p. 10.
92 J. Erlichman, 'A hungry giant looks East for fuel', *Guardian*, 17 January 1984.
93 Ibid.
94 J. Erlichman, 'ICI may turn to Russia for cheap gas supplies', *Guardian*, 6 January 1984.
95 J. Erlichman, *Guardian*, 17 January 1984.
96 S. M. Abdul-Hai and A. Al-Turki, 'Saudi basic industries corporation: programme of industrialization', in *International Symposium On Single Cell Proteins*.
97 Ibid.
98 C. Rapoport, 'ICI plans Saudi manufacturing plant', *Financial Times*, 3 January 1984.
99 J. Erlichman, 'The bug that eats gas shows Eastern promise', *Guardian*, 26 January 1984.
100 Ibid.
101 M. Moo Yung, 'A survey of SCP production facilities', *Process Biochemistry*, Vol. 11, No. 9, 1976, pp. 32–4.
102 *East–West (Fortnightly Bulletin)*, 5 June 1984, No. 340, p. 3.
103 A. Robinson, 'Food trade escapes sanctions', *Financial Times*, 22 December 1983.
104 J. Erlichman, *Guardian*, 17 January 1984.
105 S. Colvin, 'When Britain builds in Siberia', *Soviet Weekly*, 30 July 1983.
106 J. Erlichman, *Guardian*, 17 January 1984.
107 'ICI in talks on £150 m plant for Russia', *Guardian*, 6 February 1984.
108 J. Warner, 'ICI chief in Moscow trade talks', *The Times*, 16 May 1984.
109 Ibid.
110 Ibid.
111 H. Pick, 'Chernenko unsteady at cosmonauts' ceremony', *Guardian*, 6 September 1984; also *The Times*, 6 September 1984; and *Daily Telegraph*, 7 September 1984.

112 R. V. Katrush and E. G. Mirzayanova, 'Industriya belka'; S. P. Kovalenko, 'Problema kormovogo belka'; R. Rychkov, 'What microbiology can do'.
113 R. V. Katrush and E. G. Mirzayanova, 'Industriya belka'.
114 R. Rychkov, 'What microbiology can do'.
115 V. Y. Rakovskii, *Poluchenie kormovykh drozhzhei*, Minsk, 1977.
116 Ibid.
117 *SWB*: SU/W1147/A/12, SU/W1158/A/12.
118 *SWB*: SU/W1221/A/4.
119 V. Y. Rakovskii, *Poluchenie kormovykh drozhzhei*.
120 *New Scientist*, 26 May 1983.
121 *Ekonomicheskaya Gazeta*, 1983, No. 43, p. 18.
122 M. D. Zahn, 'Soviet livestock feed in perspective' in US Congress Joint Economic Committee, *Soviet Economy in a Time of Change*, Vol. 2, US Government Printing Office, Washington DC, 1979, p. 172.
123 Ibid.
124 R. S. Rychkov, 'Aktual'nye problemy razvitiya mikrobiologicheskoi promyshlennosti', *Zhurnal Vsesoyuznogo Khimicheskogo Obshchestva im. D. I. Mendeleeva*, Vol. 27, No. 6, 1982, pp. 13–17.
125 A. K. Pavlyuchenkov, *Ekonomika proizvodstva kombikormov*, Moscow, 1982, p. 88.
126 R. S. Rychkov, 'Aktual'nye problemy razvitiya'.

CHAPTER 5

1 D. S. L'vov in Akademiya nauk SSSR, Institut ekonomiki, *Ekonomicheskie problemy povysheniya kachestva promyshlennoi produktsii*, Moscow, 1969, pp. 7–8.
2 Ibid.
3 BS4891: 1972, ('A guide to quality assurance'), p. 3.
4 BS4778: 1979, ('Glossary of terms used in quality assurance'), pp. 3, 13. The definition of 'quality level' given in this standard is 'a general indication of the extent of departure from the ideal: usually a numerical value indicating either the degree of conformity or the degree of nonconformity, especially in sampling inspection'. Although this definition is widely used in statistical quality control practice, we also consider that a quantitaive assessment of quality level can be obtained from the 'tolerance zone' for products (i.e. 'the zone of values in which a measurable characteristic is in conformity with its specification' (BS4778: 1979, p. 11)), and the 'specification tolerance' (i.e. 'the permitted variation in a process or a characteristic of an item'). This latter approach is used for comparative quality assessment purposes in this present chapter.
5 K. G. Lockyer, *Factory Management*, London, Pitman, 1969, pp. 246–7.
6 J. M. Juran and F. M. Gryna, *Quality Planning and Analysis*, New York, McGraw Hill, 1970, pp. 1–4.
7 See BS4778: 1979, p. 3. This standard also defines 'grade' as 'an indication of the degree of refinement of a material or product' (p. 9).
8 D. S. L'vov, *Ekonomicheskie problemy povysheniya kachestva*.
9 J. Berliner, *The Innovation Decision in Soviet Industry*; Cambridge, Mass, MIT Press, 1976, pp. 340–1.
10 J. Grant, 'Soviet Machine Tools; Lagging Technology and Rising Imports' in US Congress Joint Economic Committee, *Soviet Economy in a Time of Change*, Vol. 1, Washington, DC, US Government Printing Office, 1979, pp. 554–80.
11 V. G. Treml, 'The inferior quality of Soviet machinery as reflected in export prices', *Journal of Comparative Economics*, June 1981, pp. 206–21.

12 A. C. Gorlin, 'Observations on Soviet administrative solutions: the quality problem in soft goods', *Soviet Studies*, Vol. XXXIII, No. 2, April 1981, pp. 163–81.
13 J. Grant, 'Soviet machine tools', p. 562.
14 M. J. Berry and M. R. Hill, 'Technological Level and Quality of Machine Tools and Passenger Cars', in R. Amann, J. M. Cooper and R. W. Davies (eds), *The Technological Level of Soviet Industry*, New Haven and London, Yale University Press 1977, pp. 523–63, but especially pp. 523–30.
15 J. Grant, 'Soviet machine tools', p. 561.
16 M. J. Berry and M. R. Hill, 'Technological level and quality', pp. 523–63, but particularly pp. 530–56.
17 Ibid., particularly pp. 547–50 and 561–3.
18 Grant considers these types of machine tools as 'conventional', but it is the author's opinion that they are better thought of as 'second generation' machines, with general-purpose machine tools as 'first generation' machines and numerically controlled machine tools as 'third generation' machines.
19 The results of this survey are published in P. Hanson and M. R. Hill, 'Soviet assimilation of Western technology: a survey of UK exporters' experience' in US Congress Joint Economic Committee, *Soviet Economy in a Time of Change*, Vol. 2, Washington DC, US Government Printing Office, 1979, pp. 582–604. This publication is a summary of the information available from the case studies; a full description of the machine-tool studies is available in M. R. Hill, *East–West Trade, Industrial Co-operation and Technology Transfer*, Aldershot, Gower Press, 1983, pp. 49–74.
20 V. G. Treml, 'The inferior quality of Soviet machinery'.
21 I. S. Shapiro, *Smetnyi spravochnik po teplomekhanicheskom oborudovaniya elektricheskikh stantsii*, Moscow, 1968, p. 105 and 1977, p. 97. The 1976 rates, published in 1977, vary from a minimum of 21 per cent for gearcutting machines to 38 per cent for lathes for 'general definition' exports; and 38 per cent for gearcutting machines to 55 per cent for drilling and boring machines for 'tropical destination' exports. The author is grateful to Professor Treml of Duke University, Durham, NC for the provision of this data.
22 V. G. Treml, 'The inferior quality of Soviet machinery'.
23 See data presented in M. R. Hill, *East–West Trade*, p. 26.
24 See, for example, 'T. I. Chesterfield limited' in *Export Dynamics Case Studies*, British Overseas Trade Board Conference on 'Breaking the Export Profitability Barriers' at the Playhouse Theatre, Nottingham, March 1977, for examples of difficulties presented by technical barriers to export of certain British products to certain West European markets, and P. G. de Monthoux, *A Note on Standards and Industrial Marketing*, Discussion Paper 77–4, International Institute of Management, Berlin, September 1977.
25 Treml quotes *Veckans affarer*, No. 35, 10 October 1974, p. 62 for evidence on Swedish costs, and a small survey of American manufacturers carried out by himself.
26 See discussion of accessories offered by Soviet, compared with Western, manufacturers in export markets in a machine-tool 'round table' discussion in *Ekomika i Organizatsiya Promyshlennogo Proizvodstva*, 1982, No. 1, p. 59.
27 J. B. Chasin and E. D. Jaffe, 'Industrial Buyers' Attitudes towards Goods made in Eastern Europe', *Columbia Journal of World Business*, Summer 1979, pp. 74–81.
28 Since Soviet market shares for engineering products in Western markets have generally been lower than their market shares in other world regions. (See M. R. Hill, *East–West Trade*, p. 26.

29 See C. D. Woodward, (ed.) *Standards for Industry*, London, Heinemann, 1965, for a collection of articles on standardization practice in a mixed economy.
30 Each Soviet state standard (GOST) specification contains the statement 'non-observance of these requirements is against the law.' In the RSFSR, the continued manufacture by any enterprise of articles having a quality lower than that specified in the relevant state standard can lead to the dismissal of the director, chief engineer and chief quality controller, or their sentence to one year's forced (or directed?) labour, or up to three years loss of freedom. (Zakony RSFSR, Postanovleniya Verkhovnogo Soveta RSFSR, 1960, p. 118, quoted in B. G. Andreev, *Ekonomicheskoe znachenie povysheniya kachestva produktsii*, Leningrad, 1968, p. 180.) Furthermore, *Ekonomicheskaya Gazeta* prints frequent accounts of enterprise profits obtained from the sale of sub-standard products being confiscated by the state through the local inspection organizations of the State Committee of Standards. M. C. Spechler ('Decentralizing the Soviet economy: legal regulation of price and quality', *Soviet Studies*, Vol. XXII, No. 2, pp. 222–54) also gives a comprehensive account of contract law between enterprises, and the role of State Arbitrazh. Spechler's paper includes accounts of several arbitration court decisions, some of which illustrate the difficulties in using comprehensive state standards to define the quality levels of some types of consumer products.
31 *Pravda*, 3 October 1965.
32 Decree of the Council of Ministers of the USSR, No. 16, January 1965; Decree of the Central Committee of the Communist Party of the Soviet Union and the Council of Ministers of the USSR, No. 729, October 1965.
33 V. V. Boitsov, *Standartizatsiya v narodnom khozyaistvo SSSR*, Moscow, 1967, pp. 270–2.
34 *Standarty i Kachestvo*, 1970, No. 12, pp. 3–6.
35 V. V. Tkachenko, *Metodika i praktika standartizatsii*, Moscow, 1967, p. 190; GOST 1.1–68, pp. 6, 7.
36 GOST 1.1–68, p. 10; GOST 1.4–68, p. 2.
37 *Standarty i Kachestvo*, 1967, No. 6, p. 70.
38 Private Communication, February 1985.
39 For a description of the procedures used in quality attestation, see the following Soviet publications:
A. A. Kokhtev, *Osnovy standartizatsii v mashinostroenii*, Moscow, 1973, pp. 124–6; F. R. Maev, *Standarty i Kachestvo*, 1977, No. 4, pp. 14–17; M. G. Lapusta and P. N. Nikitin, *Stimulirovanie povysheniya kachestva produktsii*, Moscow, 1980, p. 21; M. A. Ushakov, *Standarty i Kachestvo*, 1983, No. 12, p. 9, and I. Isaev, *Planovoe Khozyaistvo*, 1983, No. 12, p. 16.
40 See Gosudarstvennyi Komitet SSSR po Standartam, *Attestatsiya promyshlennoi produktsii po dvum kategorii kachestva*, Moscow, 1984.
41 Ibid. and Private Communication, February 1985.
42 The majority of these tests were originally developed by the German engineer, G. Schlesinger (*Testing Machine Tools*, London, Machinery, 1966) and the French engineer P. Salmon (*Machines – Outils, Reception Verification*, Paris, H. Francois et fils, 4th edn), and subsequently modified and adopted by individual companies, certain national standards organizations (including the Soviet All-Union Committee of Standardization from 1940 onwards, and the British Standards Institution from 1970 onwards) and the International Standards Organization. They are frequently referred to as alignment tests, or geometric tests, and include specifications of the tests, to be carried out, and maximum tolerances of alignment error for

each test. In addition, accuracy requirements for a sample workpiece are also specified.
43 See M. R. Hill, *Standardisation Policy and Practice in the Soviet Machine Tool Industry*, unpublished PhD thesis, University of Birmingham, 1970.
44 In 1965, for example, turning machines, including turret and capstan lathes, accounted for almost 30 per cent of total Soviet machine-tool output for that year, while milling machines accounted for 12 per cent of total Soviet machine-tool output (see N. M. Oznobin et al. (eds), *Sovershenstvovanie struktury promyshlennogo proizvodstva*, 1968, p. 136, quoted in M. J. Berry, *Research, Development and Innovation in the Soviet Machine Tool Industry*, unpublished research report, Centre for Russian and East European Studies, University of Birmingham, 1974, pp. B7–B9. The 400 mm swing 1K62 centre lathe selected for comparison, and its variants, were produced in quantities of 13,000 per year (i.e. more than 50 per cent of the total turning machine output, and hence some 12–15 per cent of the total Soviet output in 1965, using Oznobin's previously cited proportions combined with a total 1965 output figure of 186,130) (*Narodnoe khozyaistvo SSSR v 1968 godu*, 1969, p. 257). The output of the 6M82 range, also chosen for comparison, was more difficult to estimate, however. Production planning data quoted in V. A. Anufriev et al., *Krupnoseriinoe proizvodstvo frezernykh stankov*, 1965, suggest that a total of 10 machines of the 6M82 and 6M83 (1,600×400 mm table-sized machines) were produced daily (i.e. 3,000 machines annually).
45 The only British Standards relating to machine-tool accuracy which were published at that time were BS3800, 1964, which specified methods for testing the accuracy of machine tools, and a set of four standards specifying the accuracy requirements of gear-cutting machines (BS1498, 1954; BS3013, 1958; BS3329, 1961; BS3538, 1962).
46 See M. J. Berry and M. R. Hill, 'Technological level and quality', pp. 530–63.
47 M. R. Hill, *Soviet Product Quality and Soviet State Standards – The Case of Machine Tools*, Working Paper Number 87, Department of Management Studies, Loughborough University of Technology, 1984.
48 W. Seifert, *Generator-Motor: Physical Fundamentals and Basic Mechanical Forms*, Berlin, Siemens, 1976, p. 65; F. T. Bartho, *Industrial Electric Motors and Control Gear*, London, Macdonald, 1965, p. 70.
49 When the power factor is close to unity, the circuit cables can supply a correspondingly large and useful load without overheating.
50 For an explanation of how best to match efficiency and power factor to a particular application see F. T. Bartho, *Industrial Electric Motors*, pp. 48–57.
51 Ibid., p. 72.
52 See A. S. Coker, *Electric Motors*, London, Newnes, 1976, p. 29.
53 The information on Soviet-built motors was obtained from:
GOST 183–74, which covers general requirements for electrical rotating machinery; GOST 10799–77, which covers general specifications for capacitor-type squirrel cage motors; GOST 5.4–67, which covers 'mark of quality' requirements for A2–10 series squirrel cage motors; GOST 5.618–73 which covers 'mark of quality' requirements for AB2 and AB3 series squirrel cage motors; GOST 23131–78, which covers general technical requirements for A3 series motors.
Information on Western-built motors was obtained from BS4999, which covers general technical requirements for rotating electrical machinery; BS5000 Part II, which covers small power electric motors and generators; and ANSI/NEMA MG1–1978 ('Motors and Generators', published by National Electrical Manufacturers of America), and catalogues published by the following manufacturers:

Brook Crompton Parkinson Motors, Elecktrim, J. H. Fenner & Co., GEC Machines Ltd, Eroy Somer Electric Motors Ltd, Parvalux Electric Motors Ltd, Renold (Power Transmissions) Ltd.
54 R. McKay, *The Quality Levels of Asynchronous Electric Motors in the USSR*, Working Paper No. 93, Department of Management Studies, Loughborough University of Technology, August 1984.
55 GOST 5.618–73.
56 Climatic application Y and Article Category 3.
57 The general Soviet state standard on 'Application for Different Climatic Regions' (GOST 15150–69).
58 See R. McKay, *The Quality Levels of Asynchronous Electric Motors*, Appendix B.
59 See GOST 10799–77 (Squirrel cage capacitor-type motors), and BS5000 Part II (small power electric motors and generators).
60 See N. P. Ermolin and I. P. Zherikhin, *Nadezhnost' electricheskikh mashin*, Moscow, 1976, p. 118.
61 J. M. Cooper, 'Is there a Technological Gap between East and West?', paper presented at a conference on 'East–West Economic Relationships in a Changing World Economy', Canadian Institute of International Affairs, Toronto, June 1984.
62 See N. P. Ermolin and I. P. Zherikhin, *Nadezhnost*, p. 151.
63 See, for example BS5750: 1979 ('Quality Systems') for certification procedures followed by the British Standards Institution.
64 See, for example, a complaint in *Pravda*, 7 May 1983 that state standards sometimes appear to introduce minor, but unnecessary changes; and a further complaint in *Pravda*, 31 January 1985 that standards are frequently presented in an inconvenient fashion, causing a lot of extra work for the designer.
65 M. R. Hill, *The Soviet Quality Attestation and 'Mark of Quality' System*, Working Paper 84, Department of Management Studies, Loughborough University of Technology, April 1984. See also ref. 40 above.
66 See background discussion to this in J. Berliner, *passim*; R. Amann, J. M. Cooper and R. W. Davies, *passim*, and R. Amann, J. M. Cooper (eds), *Industrial Innovation in the Soviet Union*, London and New Haven, Yale University Press, 1982.

CHAPTER 6

1 P. Hanson, *Trade and Technology in Soviet–Western Relations*, New York, Columbia University Press, 1981, p. 14.
2 See, for example, J. Berliner, *The Innovation Decision in Soviet Industry*, Cambridge, Mass, MIT Press, 1978; R. Amann, J. M. Cooper and R. W. Davies (eds), *The Technological Level of Soviet Industry*, London and New Haven, Yale University Press, 1977.
3 See E. Zaleski and H. Wienert, *Technology Transfer Between East and West*, Paris, Organization for Economic Co-operation and Development, 1980.
4 'Quantification of Western Exports of High Technology Products to Communist Countries'. The reports were prepared most recently by Dr J. A. Martens and are available from the Office of Trade and Investment Analysis, US Department of Commerce. The discussion and tables in this section are based upon their most recent analysis.
5 Representing the USA, Canada, Japan, Belgium–Luxembourg, France, Federal Republic of Germany, Italy, Netherlands, Austria, Norway, Sweden, Switzerland, UK, Denmark, Finland, Ireland.
6 For example, see C. P. Ailes and A. E. Pardee Jr, *Cooperation in Science and*

Technology: an Evaluation of the US–Soviet Agreement, Boulder, Colorado, Westview Press, 1985; also, see *Review of the US–USSR Cooperative Agreements on Science and Technology*, Special Oversight Report No. 6, Subcommittee on Domestic and International Scientific Planning and Analysis, Committee on Science and Technology, US House of Representatives, 96th Congress, 2nd Session, Washington DC, US Government Printing Office, November 1976; and *Review of the US/USSR Agreement on Cooperation in the Fields of Science and Technology*, Board on International Scientific Exchange, Commission on International Relations, National Research Council, National Academy of Sciences, May 1977.

7 C. T. Saunders (ed.), *East–West Technological Cooperation: Industrial Policies and Technology Transfers Between East and West*, Vienna, Springer–Verlag, 1977; and F. Levcik and J. Stankovsky, *Industrielle Kooperation Zwischen Ost und West*, Vienna, Springer–Verlag, 1977.

8 For a fuller discussion, see Office of Technology Assessment, US Congress, *Technology and East–West Trade*, Washington DC, US Government Printing Office, 1979, pp. 102–5; and E. Zaleski and H. Wienert, *Technology Transfer*, pp. 93–138.

9 See the studies review in E. Zaleski and H. Wienert, *Technology Transfer* and George Holliday, 'Survey of sectoral case studies', *East–West Technology Transfer*, Paris, Organization for Economic Co-operation and Development, 1984, pp. 55–94.

10 CIA, *Soviet Acquisition of Western Technology*, April 1982, p. 5. Reprinted in G. Bertsch and J. McIntyre (eds), *National Security and Technology Transfer: The Strategic Dimensions of East–West Trade*, Boulder, Colorado, Westview, 1983, pp. 92–112.

11 US Congress, *Transfer of US High Technology to the Soviet Union and Soviet Bloc Nations*, Hearings before the Permanent Subcommittee on Investigations of the Committee on Governmental Affairs, US Senate, 97th Congress, 2nd Session, Washington, DC, US Government Printing Office, May 1982, p. 236–7.

12 See the report published by the US Senate (by the same Committee and under the same title as ref. 11) on 15 November 1982. Also, see L. Melvern et al., *Technobandits*, Boston, Houghton Mifflin, 1984.

13 Cited by D. Buchan, 'Western Security and Economic Strategy Towards the East,' *Adelphi Paper*, No. 192, London, International Institute for Strategic Studies, 1984.

14 CIA, *Soviet Acquisition of Western Technology*, p. 5.

15 Ibid., p. 3.

16 These studies have been reviewed and summarized by S. A. Gomulka and A. Nove, 'Contribution to Eastern growth: an econometric evaluation,' *East–West Technology Transfer*, Paris, Organization for Economic Co-operation and Development, 1984, pp. 11–51.

17 See, for example, chapter 10 in his book *Trade and Technology in Soviet–Western Relations*, pp. 161–86.

18 Ibid., 181–2.

19 G. Holliday, 'Survey of sectoral case studies,' p. 82.

20 Ibid.

21 Ibid., pp. 82–3.

22 Ibid., p. 83. Holliday argues that the discrepancy can be explained largely by the different approaches of the studies. The sectoral studies focus on the contribution in sectors which have been the primary recipients of Western technology. The country studies, on the other hand, use a macro-economic approach and examine the impact of Western technology on the total economic activity of the country.

23 Ibid.

24 *The Wall Street Journal*, 12 January 1982.
25 Department of Defense, *Soviet Military Capabilities*, Washington, DC, US Government Printing Office, 1981, pp. 71–81.
26 CIA, *Soviet Acquisition of Western Technology*, p. 10.
27 Ibid.
28 This paragraph is based upon a report published on p. 1 of the *Financial Times*, 30 March 1985.
29 J. Cooper, 'Western technology and the Soviet defence industry', in B. Parrott (ed.), *Trade, Technology, and Soviet–American Relations*, Bloomington, Indiana, Indiana University Press, 1985.
30 T. Gustafson, *Selling the Russians the Rope? Soviet Technology Policy and US Export Controls*, Rand R-2649-ARPA, Santa Monica, April 1981, p. 5.
31 G. F. Kennan, *Memoirs, 1925–50*, Boston, Little, Brown, 1967, pp. 268–9.
32 *Congressional Record*, 80th Congress, 1st session, Vol. 93, Part 6, 20 June 1947, pp. H 7493, H 7497.
33 *Foreign Relations of the United States: 1948*, 16 March 1948, Vol. IV, pp. 524–6.
34 See G. Bertsch, 'US export controls', *Journal of World Trade Law*, Vol. 15, No. 1, 1981, pp. 68–9.
35 See G. Bertsch, 'The American politics of US–Soviet trade', in B. Parrott (ed.), *Trade, Technology and Soviet–American Relations*, Bloomington, Indiana, Indiana University Press, 1985.
36 See H. Feis, *Churchill, Roosevelt and Stalin*, Princeton, Princeton University Press, 1957, p. 646.
37 See B. Jentleson, *Pipeline Politics: the Complex Political Economy of East–West Trade*, chapter 5, p. 1, unpublished manuscript.
38 H. Kissinger, *The White House Years*, Boston, Little, Brown, 1979, p. 840.
39 S. Huntington, 'Trade, technology, and leverage: economic diplomacy', *Foreign Policy*, 32, Fall 1978, pp. 64–5.
40 Bureau of Public Affairs, US Department of State, *GIST*, August 1984.
41 See H. Moyer Jr and L. Mabry, 'Export controls as instruments of foreign policy: the history, legal issues and policy lessons of three recent cases, *Law and Policy in International Business*, Vol. 15, No. 2, 1983, pp. 1–171.
42 For the early post-war period, see G. Adler-Karlsson, *Western Economic Warfare, 1946–1967*, Stockholm, Almquist and Wiksell, 1968.
43 Report by the National Security Council, 'Control of Exports to the USSR and Eastern Europe', 17 December 1947, *Foreign Relations of the United States: 1948*, Vol. IV, pp. 511–12.
44 US Department of Commerce, Fourth Quarterly Report under the Export Control Act, p. 13.
45 US Code 50 App. 2401 *et seq.*, Public Law 91-184, 83 Stat. 84.
46 For fuller detail, see G. Bertsch, *East–West Strategic Trade, COCOM and the Atlantic Alliance*, Paris, The Atlantic Institute for International Affairs, 1983.
47 US Department of Defense, *An Analysis of Export Control of US Technology*, Washington DC, Office of the Director of Defense Research and Engineering, 1976.
48 For the reaction of the scientific community, see National Academy of Sciences, Committee on Science, Engineering and Public Policy, *Scientific Communication and National Security*, 2 vols, Washington, DC, National Academy Press, 1982.
49 For an outline of Department of Defense thinking on this and related issues, see: US Department of Defense, *The Technology Transfer Control Program*, Washington DC, Department of Defense, 1984. For a wide variety of views on the issues

addressed in the technology-control section of this chapter, see M. Czinkota (ed.), *Export Controls: Building Reasonable Commercial Ties with Political Adversaries*, New York, Praeger, 1984.
50 'Singapore in COCOM hi-tech pledge', *Financial Times*, 25 April 1985.

CHAPTER 7

1 This view focuses on the inherently passive role of foreign trade in the CMEA in contrast to the market economies. Comecon's *raison d'être* is not the maximization of mutual exports (as it is in the EEC) but the assurance of stable supplies of energy and raw materials. Even when pursuing ostensibly active export policies, the exporters of manufactures perceive such exports as payments for the necessary imports of energy. Hence CMEA is conceptualized as an 'international protection system' in contradistinction to 'international trade system' such as the EEC (see V. Sobell, *The Red Market: Industrial Co-operation and Specialisation in Comecon*, Aldershot, Gower, 1984).
2 R. Amann and J. M. Cooper (eds), *Industrial Innovation in the Soviet Union*, London and New Haven, Yale University Press, 1982; or J. Berliner, *The Innovation Decision in Soviet Industry*, Cambridge, Mass, MIT Press, 1976.
3 F. D. Holzman, 'Some systemic factors contributing to convertible currency shortages of CPEs', *American Economic Review*, Vol. 69, No. 2, p. 78.
4 P. Hanson, 'The Soviet system as a recipient of foreign technology', in R. Amann and J. M. Cooper, *Industrial Innovation*, p. 416.
5 P. Hanson, *Trade and Technology in Soviet–Western Relations*, London and Basingstoke, Macmillan, 1981, p. 14.
6 S. Gomulka, *Inventive Activity, Diffusion and the Stages of Economic Growth*, Economics Institute, Aarhus University, Monograph 24, 1971.
7 The share of capital investment allocated to reconstruction and modernization (as opposed to 'green field' construction) now varies between 50 and 70 per cent of the total (L. Nartsissov and A. Shuianska (CMEA Secretariat), *Ekonomicheskoe sotrudnichestvo stran chlenov SEV*, 1984, No. 5, p. 18).
8 Bulgaria had approximately the same share as the USA; Poland's share was larger than Britain's; Hungary had larger shares than Italy, France or Austria; only the USSR and Romania had relatively low shares but still higher than the Benelux Countries (see Z. M. Fallenbuchl, 'The commodity composition of intra-Comecon trade and the industrial structure of the member countries', in *Comecon: Progress and Prospects*, NATO Colloquium, NATO Directorate of Economic Affairs, Brussels, 1977, pp. 107–8).
9 UN Economic Commission for Europe, *Analytical Report on the State of Intra-European Trade*, New York, 1970.
10 'Specialized exports' denote exports conducted under the terms of co-operation and specialization agreements concluded among the CMEA countries.
11 A. Chaushev, *Ekonomicheskoe Sotrudnichestvo Stran Chlenov SEV*, 1983, No. 8, p. 45; engineering occupied over 80 per cent of total turnover in 'specialized trade'.
12 Doubts to this effect are frequently aired in the Eastern European literature.
13 In recent years an interesting discussion has developed on the meaning and extent of Soviet subsidization of Eastern Europe (see J. M. van Brabant, 'The USSR and socialist economic integration – a comment', *Soviet Studies*, Vol. XXXVI, No. 1,

January 1984; M. Marrese and J. Vanous, *Soviet Subsidization of Trade with Eastern Europe – A Soviet Perspective*, Berkeley, University of California, 1983; and R. Dietz, *Advantages/Disadvantages in USSR Trade with Eastern Europe – the Aspect of Prices*, WIIW, Vienna, 1984).

14 V. Sobell, 'The CMEA at the close of the century', *Radio Free Europe, RAD Background Report/8*, 4 February 1985.
15 O. Bogomolov, *Kommunist*, 1983, No. 7.
16 V. Sobell, *The Red Market: Industrial Co-operation and Specialisation in Comecon*; L. E. Nolting, *Integration of Science and Technology in CMEA*, Washington DC, Bureau of the Census, Foreign Economic Report No. 21, 1983.
17 See Yu. Masliukov (Chairman of the Intergovernmental Commission for Co-operation in Computer Technology) *Ekonomicheskoe Sotrudnichestvo Stran Chlenov SEV*, 1984, No. 1, pp. 15–17.
18 I. Sebestyen (CMEA Secretariat), *Ekonomicheskoe Sotrudnichestvo Stran Chlenov SEV*, 1984, No. 2, p. 31.
19 See, for example, the interview with the Director of the GDR association 'Robotron', F. Vokurka, in *Ekonomicheskoe Sotrudnichestvo Stran Chlenov SEV*, 1983, No. 10, p. 52; for an example of GDR–CSSR–Western co-operation see *Prace*, 13 March 1984, p. 4.
20 L. Suran, 'The development of microprocessors in Czechoslovakia – the components base of automation', *Automatizace*, 1984, No. 5, pp. 115–19.
21 In the wake of the preparation for intensified co-operation in engineering for the 1986–90 plan and beyond, the former Standing Commission for Machine-Building was upgraded into the CMEA Committee for Co-operation in Machine-Building. The first session of the Committee took place in Moscow in February 1985.
22 A notable Western effort is in J. M. van Brabant, *Socialist Economic Integration: Aspects of Contemporary Economic Problems in Eastern Europe*, Cambridge, Cambridge University Press, 1980, pp. 195–206.
23 At the time of writing, expectations to this effect gained currency with the accession of M. Gorbachev as the top Soviet leader.

CHAPTER 8

1 P. Hanson, 'Success indicators revisited: the July 1979 Soviet decree on planning and management', *Soviet Studies*, Vol. XXXV, No. 1, January 1983.
2 N. Nagibin and I. Ryazhskikh, 'Eshche raz ob aksaiskom metode', *Pravda*, 18 January 1975, p. 2.
3 S. D. Reznik, 'Trudovaya distsiplina', *Ekonomika Stroitel'stva*, 1980, No. 3.
4 'O dal'neishem ukreplenii trudovoi distsipliny i sokrashchenii tekuchesti kadrov v narodnom khozyaistve', *Ekonomicheskaya Gazeta*, 1980, No. 3, p. 4.
5 V. Fil'ev, 'Shchekinskii metod i perspektivy ego dal'neishego razvitiya', *Voprosy Ekonomiki*, 1983, No. 2.
6 G. Kulagin, 'Trudno byt' universalom', *Pravda*, 8 December 1982, p. 2.
7 Yu. Andropov, 'Uchenie Karla Marksa i nekotorye voprosy sotsialisticheskogo stroitel'stva v SSSR', *Voprosy Ekonomiki*, 1983, No. 3.
8 G. Aliev, Speech on the draft law on working collectives, *Pravda*, 18 June 1983, p. 2.
9 I. Samoshchenko, 'Kollektiv v sisteme upravleniya', *Ekonomicheskaya Gazeta*, 1983, No. 35, p. 10.
10 T. Zaslavskaya, *Doklad o Neobkhodimosti Bolee Uglublennogo Izucheniya v SSSR*

References

   Sotsial'nogo Mekhanizma Razvitiya Ekonomiki, Radio Liberty, Materialy Samizdata, No. 35/83, 26 August, AC no. 5042.
11 'V TsK KPSS, Sovete Ministrov SSSR i VTsSPS', Ekonomicheskaya Gazeta, 1983, No. 33, pp. 3–4.
12 K. Chernenko, speech reported in Ekonomicheskaya Gazeta, 1984, No. 16.
13 'V Tsentral'nom Komitete KPSS i Sovete Ministrov SSSR', Ekonomicheskaya Gazeta, 1983, No. 31, p. 5.
14 I. Koval', 'Zakrepit' dostignutoe, idti dal'she', Ekonomicheskaya Gazeta, 1984, No. 31, p. 6.
15 I. Goncharov, 'Sebestoimost' – pokazatel' opredelyayushchii', Ekonomicheskaya Gazeta, 1984, No. 10, p. 8.
16 I. Koval', 'Zakrepit' dostignutoe'.
17 B. Ural'tsev, 'Otvetstvennost' vo vsekh zven'yakh', Ekonomicheskaya Gazeta, 1984, No. 35, p. 8.
18 S. Tsagaraev, 'Eksperiment i vstrechnyi', Ekonomicheskaya Gazeta, 1984, No. 28, p. 8.
19 B. Ural'tsev, 'Otvetstvennost' vo vsekh zven'yakh'.
20 T. Timonina, 'Ot ritmichnosti proizvodstva – k ritmichnosti postavok', Ekonomicheskaya Gazeta, 1984, No. 32, p. 6.
21 K. Nedoguiko, 'Stimuly ritmichnoi raboty', Ekonomicheskaya Gazeta, 1984, No. 27, p. 6.
22 A. Matulyavichus, 'Kak trebuet eksperiment', Ekonomicheskaya Gazeta, 1984, No. 21, p. 13.
23 A. Bionchuk, 'Mnogoe prishlos' menyat'...', Ekonomicheskaya Gazeta, 1985, No. 8, p. 7.
24 V. Kazakov, 'Eksperiment – eto poisk', Ekonomicheskaya Gazeta, 1984, No. 24, p. 7.
25 G. Pakhomov, 'Eksperiment stavit zadachi', Ekonomicheskaya Gazeta, 1984, No. 34, p. 8.
26 S. Tsagaraev, 'Eksperiment i vstrechnyi'.
27 A. Gnidenko and I. Divnogortsev, 'Bol'she prav – vyshe otvetstvennost'", Ekonomicheskaya Gazeta, 1984, No. 30, p. 8.
28 M. Rudoi, 'K eksperimentu gotovy', Ekonomicheskaya Gazeta, 1984, No. 51, p. 8.
29 I. Koval', 'Zakrepit' dostignutoe'.
30 B. Ural'tsev, 'Otvetstvennost' vo vsekh zven'yakh'.
31 V. Velichko, 'Eksperiment otkryvaet novye gorizonty', Ekonomicheskaya Gazeta, 1985, No. 6, p. 6.
32 A. Gnidenko and I. Divnogortsev, 'Bol'she prav – vyshe otvetstvennost'".
33 A. Shukhov, 'V rezhime eksperimenta', Ekonomicheskaya Gazeta, 1984, No. 20, p. 9.
34 V. Vostrikov, 'S novoi merkoi k otsenke raboty kazhdogo', Ekonomicheskaya Gazeta, 1985, No. 14, p. 13.
35 N. Tikhonov, 'Za protsvetanie sovetskoi rodiny', Pravda, 2 March 1984, p. 2.
36 M. Gorbachev, speech reported in Ekonomicheskaya Gazeta, 1985, No. 16, p. 3.
37 'V Tsentral'nom Komitete KPSS i Sovete Ministrov SSSR', Ekonomicheskaya Gazeta, 1983, No. 36, p. 5.
38 Report of conference proceedings, Ekonomicheskaya Gazeta, 1985, No. 16, p. 4.
39 A. Tselikov, 'Metall i mashiny', Pravda, 6 March 1980, p. 2.
40 Economist Intelligence Unit, Quarterly Economic Review of the USSR, 1984, No. 3, pp. 11–12.
41 Economic Commission for Europe, Economic Survey of Europe in 1983, New York, United Nations, 1984, table 3.4.1.
42 'V Tsentral'nom Komitete KPSS i Sovete Ministrov SSSR', Ekonomicheskaya Gazeta, 1984, No. 23, pp. 6–7.

43 A. Deminov, 'Novye smetnye normy i tseny', *Ekonomicheskaya Gazeta*, 1984, No. 15, p. 8.
44 S. Bulkagov, 'Metodologicheskaya osnova i problemy sozdaniya edinoi sistemy planirovaniya kapital'nogo stroitel'stva', *Ekonomika Stroitel'stva*, 1984, No. 10, p. 12.
45 A. Klyuev et al., 'Sovershenstvovanie sistemy khozraschetnykh vzaimootnoshenii i provedenie shirokomasshtabnogo eksperimenta v sfere kapital'nogo stroitel'stva', *Ekonomika Stroitel'stva*, 1984, No. 7.
46 N. Tikhonov, 'Za protsvetanie sovetskoi rodiny'.
47 A. Berezin, 'Vospitanie kollektivizma', *Ekonomicheskaya Gazeta*, 1984, No. 15, p. 5.
48 N. Fedorenko, 'Ekonomiko-matematicheskie modeli i metody', *Ekonomicheskaya Gazeta*, 1985, No. 1, p. 14.
49 D. A. Dyker, *The Future of the Soviet Economic Planning System*, London, Croom Helm, 1985, table 2.3.

CHAPTER 9

1 See, for example, A. Bergson, 'Technological progress', in A. Bergson and H. S. Levine, *The Soviet Economy Toward the Year 2000*, London, George Allen & Unwin, 1983, pp. 34–78.
2 For a more complete discussion see H. Levine, D. Bond, C. Movit and E. Goldstein, *The Causes and Implications of the Sharp Deterioration in Soviet Economic Performance*, Washington, DC, Wharton Econometric Forecasting Associates, Inc., 1983. Also, H. S. Levine, 'Possible causes of the deterioration of Soviet productivity growth in the period 1976–1980', in US Congress Joint Economic Committee, *Soviet Economy in the 1980s: Problems and Prospects*, Part I, Washington, DC, US Government Printing Office, 1982, pp. 153–68.
3 See H. Hunter and D. Kaple, 'Transport in trouble', in US Congress Joint Economic Committee, *Soviet Economy in the 1980s*, pp. 216–41.
4 See the chapter by C. Movit in H. Levine, et al., *The Causes and Implications*.
5 *USSR: Measures of Economic Growth and Development, 1950–80*, Studies prepared for the use of the Joint Economic Committee, US Congress, Washington, DC, US Government Printing Office, 1982.
6 See E. Hewett, 'Foreign economic relations' in A. Bergson and H. S. Levine, *The Soviet Economy*, pp. 269–310.
7 See the chapter by E. Goldstein in H. Levine et al., *The Causes and Implications*.
8 See J. Vanous, 'Recent developments in Soviet gross hard-currency debt and assets', *CPE Current Analysis*, Vol. IV, No. 18, 19 March 1984, Washington, DC, Wharton Econometric Forecasting Associates, Inc., 1984.
9 See D. L. Bond and H. S. Levine, 'The Soviet domestic economy in the 1980s', in H. Sonnenfeldt (ed.), *Soviet Politics in the 1980s*, London, Westview Press, 1985, pp. 67–84.
10 The following analysis and forecast for the Soviet energy sector is taken from *Centrally Planned Economies Longterm Projections*, Washington, DC, Wharton Econometric Forecasting Associates, Inc., 1985.
11 *Centrally Planned Economies Outlook*, Vol. 6, No. 1, April 1985, Washington, DC; Wharton Econometric Forecasting Associates, Inc., 1985.
12 For example, see G. Schroeder, 'Soviet economic "reform" decrees: more steps on the treadmill', in US Congress Joint Economic Committee, *Soviet Economy in the 1980s*, pp. 65–88.
13 N. Nimitz, 'Reform and technological innovation in the 11th Five-Year Plan', in

S. Bialer and T. Gustafson, *Russia at the Crossroads: the 26th Congress of the CPSU*, London, George Allen & Unwin, 1982, pp. 140–55.
14 Ibid., p. 152.
15 Chapter by D. A. Dyker in this volume. See also M. Bornstein, 'Improving the Soviet economic mechanism', *Soviet Studies*, Vol. XXXVII, No. 1, January 1985, pp. 1–30.

Index

Academy of Sciences, 17–18
Afghanistan, 120, 128, 130, 131
agriculture, 65, 75, 153, 155, 157, 175
Aliev, G., 156
Amann, R., 180
Andropov, Yu V., 6, 7, 154, 155, 156, 158, 160, 175
Aviation Industry, Ministry of the, 42, 44, 126

Balashov, E. P., 54
Bal'mont, B., 59
Baltic Republics, 160
Becker, A., 1
Belorussia, 160, 164, 166
Bergson, A., 8
Berliner, J., 95
Berry, M. J., 96
Bialer, S., 1
biotechnology, 75–93
  single cell protein industry:
  background, 76–82; choice of raw materials, 78, 85, 86, 92 (n-paraffins, 78–82, 85; wood wastes, 85; sulphite waste liquor (SWL), 86–7); computerization, 83–5; environmental pollution, 82; hydrolysis industry, 87–8; production levels, 88–93; Soviet–ICI deal, 89–92; utilization of waste, 85
Bobrovskii, P. A., 82
Bogomolov, O., 146
Brezhnev, L. I., 5, 10, 31, 34, 39, 43, 153, 154, 155, 157, 158, 175
British Petroleum, 75
British Standards, 94, 99, 104, 106

Bucy, Fred, 132
Bulgaria, 55

Carter, President, 131
  Carter Administration, 129, 133
Chernenko, K. U., 24, 154, 158, 165, 168
CIA, 121, 122, 124, 169
COCOM, 62, 63, 127, 128, 130–34, 136
Cold War, 33
Comecon countries, technology transfer, 135–52
  consequences of centrally-planned economies, 139–141; dependence on Western technology, 147; effect of Comecon, 135–7; 'exchange of inefficiency', 137–9; intra-bloc prices, 144–5; joint development of technology, 61, 64, 146–50, 152; machinery flows, 141–4; relations between USSR and Eastern Europe, 144–6; Summit Meeting (1984), 145–6, 152, 179
Communications Equipment Industry, Ministry of the, 58
Communist Party, 58, 158
computerization, 168–9 *see also* microcomputers
construction sector, 65, 164–7, 171
Control Data Corporation, 60
Cooper, Julian, 11, 15, 126
Council of Ministers, 32, 77, 89, 99, 100
Czechoslovakia, 55

Defence Industry, Ministry of the, 32, 39, 42, 43, 44
defence sector, 2, 3, 13, 31–50, 52, 58
  civilian production (contribution to

economy, 48–9; historical background, 33–4; organization of production, 40–44; range of products manufactured, 34–9; technological level, 44–5; volume of production, 39–40)
Dmitriev, I. F., 32
Dyker, D., 114, 180

Economics of Construction, 167
Economist, The, 55
economy, Soviet, 170–81
  capital investment, 98, 171; capital-output ratios, 8; growth rates, 1, 7, 98, 170–74, 177–8; foreign trade, 8, 174; import objectives, 98; imports from West, 117, 123, 179–81; investment planning, 164–7, 175; loans to Third World, 175; planning mechanism, 1; supply uncertainty, 154, 156, 162
Ekonomicheskaya Gazeta, 59, 92
electric motors, 106–13
Electronics, 60
electronics industry, *see* microcomputers, microprocessors
Electronics Industry, Ministry of the, 40, 58, 59, 62
energy sector, 11, 177–9
Erlichman, J., 91
Ermolin, N. P., 112
Eroshin, V. K., 77
export price supplements, 97
Exxon, 90

Fallenbuchl, Z. M., 141
Fedorenko, N., 169
ferrous metals sector, 174
food processing, 65
Food Programme (1982), 175

Germany, East, 54–5, 61, 87
Germany, West, 11, 117
Gorbachev, M. S., 1, 18, 24, 163, 167, 168, 179
Gorlin, A. C., 95
Gosplan, 32, 43, 158
Gossnab, 161
Grant, J., 95
Gustafson, T., 126

Hanson, Philip, 1, 59, 116, 123, 137
Harriman, W. Averell, 129
Harvey-Jones, John, 91
Hill, M. R., 96, 102
Hohmann, H., 14, 15
Humphrey, A. E., 83, 88
Hungary, 180
Huntington, S., 129

Iceland, 131
ICI, 89–92
industrial innovation, 16–21, 51–2, 60–61, 83, 98, 113, 116, 158, 164–5
industrial planning, 158–63
Inman, Admiral B., 122
Instrumentation, Automation and Control Systems, Ministry of, 58
International Union for Pure and Applied Chemistry, 78, 82
investment planning, 164–7

Japan, 11, 131, 180
Juran, J. M., 94

Kalashnikov, M. T., 35
Kamyenni, V. I., 88
Kennan, George, 127
Kennedy Administration, 129
KGB, 126
Kiser, J., 14
Kommunist, 24
Kontorovich, V., 9
Korean Air Lines incident, 130
Kostandov, L., 91, 92
Kosygin, A., 164
Khrushchev, N., 157, 169
Kudinov, V., 17
Kuznetsov, V. Ya., 61

labour planning, 155–7
labour productivity, 7, 155–6
Lithuania, 160
Lockyer, K. G., 94
Losinov, A. B., 77
L'vov, D. S., 94, 95

machine tool industry, 96–7, 102; *see also* product quality
Malenkov, G. M., 34

Index

Marchuk, G., 17, 19
Martens, J. A., 11
Marx, Karl, 167
Maslyukov, Yu. D., 43
Melman, S., 49
Merkin, R. M., 167
Mexico, 91
microbiology, *see* biotechnology
microcomputers, 51–74
  personal computers, 11, 37, 57–8, 62–3; production in USSR, 55–7
microprocessors, 3, 51–74, 147
  application, 51–3, 60–63, 65; CMEA programme, 61, 64; development trends, 63–4; production in USSR, 53–5, 58–60; similarities between Western and Soviet machines, 54, 59–60, 62, 65
Mikhailov, V., 43
Mitsubishi, 90
Mobil, 90

Narodnoe Khozyaistvo SSSR, 9
Nimitz, N., 180
Nixon Administration, 129
Not by Bread Alone (Dudintsev), 16

Organization of Arabian Petroleum Exporting Countries (OAPEC), 91
Organization for Economic Cooperation and Development (OECD), 9

patents, 11
'perfect computationists', 154, 168–9
petrochemical industry, 82
planning reforms, 153–69
  industrial planning, 158–63; investment planning, 164–7; labour planning, 155–7, 161–2; planning bureacracy, 157; Research and Development, 163–4
Pokrovsky, A. A., 77
Poland, 120, 128, 130
Posa, John, 60
product quality, 94–114, 157–8, 160
  comparison with West, 94–7, 98, 103–6, 107, 108, 112; economic aspects, 94; electric motor manufacturing standards, 106–13; legal aspects, 95; machine tool standards, 102–6; quality attestation system, 101, 107, 113, 114
Pravda, 59
Proleiko, V. M., 60
Puzankov, D. V., 54

railway system, 171
Reagan, President, 115, 124, 128, 130, 131, 132, 133, 179
Research and Development, 10, 16, 17, 18, 19, 20, 24, 124, 163–4
Rimmington, T., 13
Rychkov, R., 82, 93
Ryumin, A. A., 40

Saudi Basic Industries Corporation, 90–91
Shcharansky, A., 129
Semushina, T. M., 88
Sharkova, V. I., 76
Shchekino system, 155
Shell, 90
Shokin, A. I., 58
Singapore, 134
single cell protein, *see* biotechnology
Skryabin, G. K., 77
Smirnov, L. V., 32
Snell, P., 12
Sobolev, Yu. A., 43
Soviet Acquisition of Western Technology (CIA), 124
Soviet Military Capabilities (US Defense Dept), 124
Speichim, 78
Stakhanovite movement, 168
Stalin, J., 1, 16
State Committee for Science and Technology, 17, 18, 61
state standards, 99–113
Stukolov, P., 59
Subbotskii, Yu. V., 39
Sweden, 97

technological progress, 5–30
  diffusion of new technologies, 10, 11–13, 15, 20; economic indicators, 7–9; exports to West, 14, 49; systemic influences, 16–21
technology transfer, 115–34, 179

composition, 116–21; controls, 115, 127–34; impact of transfers on USSR and Eastern Europe, 122–6; intangible transfers, 121–2, 133; inter-government agreements, 120; intra-Comecon, *see* Comecon; national security implications, 121, 127, 128, 130–34
Tikhonov, N., 91, 162, 127
Trapeznikov, V., 7, 19
Treml, V. G., 95, 97
Truman, President, 130

Union Carbide, 78

Velikhov, E. P., 57–8
Ventsov, S., 33

Weinberger, Caspar, 124

Young, J. P., 11

Zahn, M. D., 92
Zaslavskaya, T., 6, 157
Zherikhin, I. P., 112
Zhimerin, D., 61